PHYSICS FROM THE EDGE

EDGE

A New Cosmological Model for Inertia

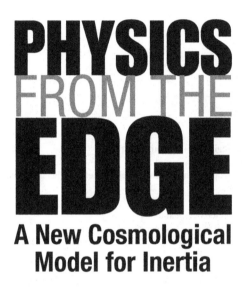

PHYSICS
FROM THE
EDGE

A New Cosmological
Model for Inertia

Michael Edward McCulloch

University of Plymouth, UK

World Scientific

NEW JERSEY • LONDON • SINGAPORE • BEIJING • SHANGHAI • HONG KONG • TAIPEI • CHENNAI

Published by

World Scientific Publishing Co. Pte. Ltd.
5 Toh Tuck Link, Singapore 596224
USA office: 27 Warren Street, Suite 401-402, Hackensack, NJ 07601
UK office: 57 Shelton Street, Covent Garden, London WC2H 9HE

Library of Congress Cataloging-in-Publication Data
McCulloch, Michael Edward, author.
    Physics from the edge : a new cosmological model for inertia / Michael Edward McCulloch,
University of Plymouth, UK.
      pages cm
    Includes bibliographical references and index.
    ISBN 978-9814596251 (hardcover : alk. paper)
    1. Inertia (Mechanics) 2. Momentum (Mechanics) 3. Acceleration (Mechanics) I. Title.
    QC137.M34 2014
    531'.12--dc23

                                                              2013051148

British Library Cataloguing-in-Publication Data
A catalogue record for this book is available from the British Library.

Typeset by Stallion Press
Email: enquiries@stallionpress.com

Printed in Singapore

To my wife

Acknowledgements

Thanks to my wife Boksun Kim for support and encouragement, to Andrew Jones, Rodney McCulloch and Jeehoon Kang for proof reading early chapters, and editors Kellye Curtis & Kah-Fee Ng for all their help.

Contents

Inertia: A Blind Spot in Physics

Feynman: Say, pop, I noticed something. When I pull my wagon, the ball rolls to the back of the wagon and when I'm pulling it along and I suddenly stop, the ball rolls to the front of the wagon. Why is that? Feynman's Dad: That nobody knows. The general principle is that things which are moving tend to keep on moving and things which are standing still tend to stand still, unless you push them hard. This tendency is called 'inertia' but nobody knows why it is true (Feynman, 1988).

Chapter 1

A History of Inertia

Aristotle

The property that we now call inertia (explained very clearly by Feynman's father in the quotation) remained hidden from view for a long time. The Greek philosopher Aristotle, for example, said that after a force is applied to objects, they tend to come to a rest after a while, unless a force is continually applied (Aristotle, 330BC, *Physics*). He thought that the natural state was one of rest. This is a natural idea to have since most things in our experience are at rest, and of course for humans the effortless state is one at rest. It takes some effort to get up and go get a beer from the fridge, so it is easy to assume nature behaves the same way. Based on this assumption (the default state is zero velocity), Aristotle argued that the Earth did not move. He said that when you throw something up into the air it lands on the same place it was thrown from and if the Earth was moving the thrown object would be left behind by the spinning Earth and land elsewhere.

Aristotle was a relatively good observer, but at this time Greek science had not progressed towards the idea of experiment, a process by which physicists find ways to purge most other influences from a system so that the influence you want to look at stands out undiluted. One example of such an experiment is Feynman's wagon, a ball rolling in a wagon feels less friction, so other effects, like inertia, are more apparent in its behaviour.

At the moment I am writing this I am watching the SpaceX Dragon spacecraft as it attempts to dock with the International Space Station, and this is a good example of an inadvertent inertial experiment since in space there is no air to cause drag so the motion of the Dragon very clearly shows that things keep going at a constant speed

2   *Physics from the Edge: A New Cosmological Model for Inertia*

Figure 1.   The accepted natural state of motion is a constant velocity. People have an intuitive understanding of this, see the cartoon above, but it was first brought into physics by Galileo Galilei.

if there is no force applied (forgetting for a moment the elliptical orbit which is due to the gravitational force pulling the Dragon towards the Earth). When Dragon fires its engines, and these long bursts of light are clearly visible in the live video, it does change the speed of the spacecraft and they have to be careful they don't coast forward too far at constant velocity and bang into the ISS. So the default state is not rest, but constant speed. Aristotle did not have access to experiments in the vacuum of space, but he could conceivably have used rolling balls.

Galileo

It was Galileo who first combined the experimental method with mathematical rigour. Galileo had been trained in music from an early age and because of that he became interested in mathematics, and this was a great advantage. Mathematics allows you to prove things in a far more objective way: it keeps you honest. The ancient Greeks tended to think that one could arrive at the truth through debate, but ordinary language is vague enough that mistakes can be hidden in nuance and definitions, and it is too easy for glib fellows to win arguments through charisma or cheap tricks.

As an example, one early argument of Galileo's time (see White, 2007, p44) involved a hail storm that occurred in Pisa shortly after he moved there. His colleagues cited Aristotle's ideas in which heavier

objects dropped from an equal height were supposed to accelerate more rapidly and hit the ground first. Galileo pointed out that this had not been observed, large and small hail stones had hit the ground at the same time. His opponents (including the then famous academic Girolamo Borro) then said that this did not contradict Aristotle since the smaller ones must have started from a lesser height. This is a very good example of something that also happens today, the *ad hoc* rearrangement of unknown quantities to save a popular theory. The clue back then was the arbitrary way Galileo's opponents had to arrange the differently-sized hailstones to agree with the observation of an equal fall time: there was no reason for it. There is similarly no reason known for the distribution of dark matter in a halo round galaxies (but more on this later).

Galileo, annoyed by the slippery nature of his opponents' arguments, used a nice thought experiment to prove that stones of different weight fall at the same speed. He first assumed that the opposite was the case, and disproved this possibility logically as follows: consider a stone A falling (see Figure 2, left). Now break it into two halves B and C (Figure 2, right). If Aristotle is right then both B and C will fall slower than A. But B and C together constitute

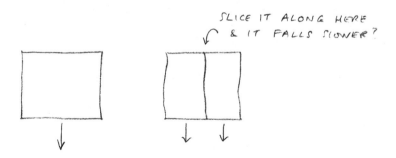

Figure 2.   Galileo's argument that different masses fall at the same rates by reductio ad absurdum. Consider first the theory of Aristotle that heavier objects fall faster. Consider a large block falling (on the left). Now cut it in two down the centre with a laser, without applying any significant external force. Suddenly you have two blocks falling of half the weight so Aristotle would predict they should suddenly fall less fast, but there has been no force applied! Galileo said the only consistent thing we can assume is that fall speed is independent of mass (of course, neglecting air resistance).

A. Therefore this implies that A will fall slower than itself! The only way to resolve this paradox is to assume that all stones fall at the same speed (see also Hoffman, 1906, p25). At the time it was difficult to prove this experimentally, since accurate clocks did not yet exist and falling objects move too fast to be easily timed. Galileo solved this problem by rolling balls down inclined planes, this reduced the apparent force of gravity since only a small component of gravity was directed down the slope, and he timed these "slow motion" experiments using a water clock. He showed that all the balls accelerated equally irrespective of their mass.

This flatly contradicted Aristotle, who had claimed that a 100 pound ball falling from 100 cubits (58 metres) height would hit the ground before a 1 pound ball could fall one cubit (White, p59). Girolamo Borro, then the most famous academic in Pisa tried this experiment and confirmed Aristotle's prediction. Galileo tried the experiment himself and disagreed (White, p59).

We now know (within the accuracy of experiments) that objects fall together irrespective of their mass. This is because although heavier objects are more attracted to the Earth due to their greater gravitational mass, they also have more inertial mass and so find it harder to accelerate (see Figure 3). These two effects cancel (although in this book I will try to convince you that things are slightly more complex). This was demonstrated elegantly, although not particularly rigorously, by the Apollo astronaut David Scott when he went to the Moon. The Moon has no atmosphere so when he dropped a hammer and a feather there they hit the lunar surface together.

Galileo also discovered the law of inertia (which can be summarised by saying that the default acceleration is zero). He tried several experiments with V shaped planes and found that if the balls were rolled down from a height H, on one side of the V, then they would rise to the height H on the other side. This is a consequence of the conservation of energy which can be written:

$$\text{Potential energy} + \text{Kinetic energy} = \text{Constant} \qquad (1.1)$$

The potential energy an object has is due to its being high up in a gravitational field, so it has the potential to move, and the kinetic energy it has is because of movement and the energy that can be released when it is stopped. The balls start with lots of potential energy (PE) because of their height, convert this potential energy to kinetic energy (KE) as they roll down one side of the V and then are able to climb up the other side of the V by converting the kinetic energy back into potential energy. Some energy is always lost in this process and becomes internal energy — i.e.: heat. Galileo asked what would happen if the second half of the V was horizontal? Then the ball would be able to keep all the kinetic energy it converted from its initial potential energy and would roll horizontally forever! In reality, the ball would eventually stop due to the energy lost to internal energy (heat) which we call friction, but Galileo realised that this friction had for millennia obscured the basic law which was that things keep going. More formally, the principle of inertia is as follows:

A moving body will continue in the same direction unless disturbed.

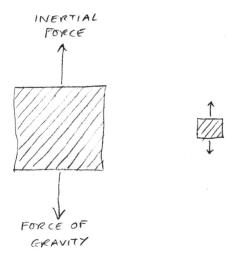

Figure 3.   **The principle of equivalence.** Objects with more mass (left) are attracted more by the gravity of the Earth, but also have more inertial mass so find it more difficult to accelerate given the same force. The assumption then is that the up and down arrows are always equal in length. This means acceleration is independent of the mass — the amount of 'stuff' in the object.

6   *Physics from the Edge: A New Cosmological Model for Inertia*

This is also true of "A ball *rolling* on a level surface..." but anyway, this is the model that is demanded by Feynman's wagon experiment and it could have been found out by the Greeks if they had had a modern kind of experimental attitude or had been keen bowls players. It shows that the appropriate question to ask is not "what makes it go" since going is the natural state, as the Greeks did, but rather "what makes it stop" (Feynman, 1985) since it is stopping that requires the input energy.

This principle of inertia is important because it defeated Aristotle's objection to the moving Earth. Although Copernicus had already proposed that the Earth moved before Galileo found his law of inertia, it had been thought that if the Earth moved then everything on it would have fallen behind to maintain their default position. Now it was understood that the important thing was not the default position, but rather the default velocity and so long as this did not change then the Earth could move along at a constant speed and happily drag trees, water and philosophers along with it. This does not yet consider acceleration, but Galileo's inertia helped to make Copernicus' hypothesis more plausible and helped the more mathematical models of it derived later by Kepler and Newton.

Galileo made this slightly more general by using the example of a sailing ship. On board a ship travelling at a constant speed the laws of dynamics work just the way they do in an unmoving reference frame, say on land, because the default state of matter is a constant velocity (zero acceleration) and not a fixed place, so one cannot tell whether one is moving by any dynamics experiment onboard the ship, but only by looking outside. Einstein later extended this to include electromagnetic experiments on the ship (this is special relativity) and accelerated motions (with general relativity).

Newton

Isaac Newton was born on Christmas day, 1642 in Woolsthorp, near Grantham in England, and he could be said to have invented modern physics since he was an expert in the use of experiment to isolate systems so they could be studied and the subsequent application of rigid mathematics, after the observations have been completed, to

Figure 4.   In an inertial (unaccelerated) reference frame, for example on a ship moving at a constant speed on a flat sea, the laws of physics are the same as they would be in a static ship, so one cannot tell you are on a moving ship if there is no way to look outside. That is the theory of Galilean relativity, but it only applies to a ship on a perfectly calm sea. As any sailor knows the reality is that surface waves make the ship accelerate, so inertial frames rarely occur.

describe the system. He was good at both occupations. In his first law of motion, Newton stated the law of inertia that he got from Galileo like this:

> *Unless acted upon by a net unbalanced force, an object will maintain a constant velocity.*

More crucially, he began to apply it mathematically to objects in space, like the Moon, where, because of the lack of friction, it is more obviously valid than on the Earth. Galileo did not make a distinction between gravitational mass and inertial mass. This was done by Newton. Richer provided a clue when he found that the period of oscillation of a pendulum could be increased by taking it from Paris to Cayenne in French Guyana (Bradley, 1971, p107). The period of a pendulum is given by

$$T = 2\pi \sqrt{\frac{L}{g}} \qquad (1.2)$$

where L is the length of the pendulum, and g is the acceleration due to gravity. At the equator, the centrifugal outwards force due to the Earth's rotation counters the acceleration due to gravity (g) so that it appears to be slightly less, so the period (T) of a pendulum is greater. The amount of mass, defined as: density × volume, in the pendulum was the same, but its dynamics (the force on it) had changed.

Since bodies with different weights (balls, feathers) have equal acceleration, then it is clear that acceleration is not just proportional to the force on something (weight) and we have to say

$$\text{Acceleration} = \text{weight (force)/something else.} \qquad (1.3)$$

This 'something else' is a property that could be called 'resistance to acceleration' and is now called inertial mass (m_i) and the formula becomes:

$$F = m_i \times a \qquad (1.4)$$

This is Newton's second law of motion, and states that when you apply a given force in newtons (N) the acceleration you get is diluted by the amount of inertial mass the body has. If you push a car with your little finger, the acceleration is so small that it won't even overcome friction, but if you push a feather with your little finger it will accelerate. Newton's first law can then be derived from this second one. If we have no external force applied, $F = 0$ and so:

$$0 = m_i \times a \qquad (1.5)$$

Therefore the acceleration is zero and we get the first law as described above. This makes it seem that all you need actually is Newton's second law, and not the first. Newton also used something called the gravitational mass. He defined the force (F) of gravity as

$$F = \frac{GMm_g}{r^2} \qquad (1.6)$$

where G is a constant approximately equal to 6.6×10^{-11} m^{3}kg^{-1}s^{-2}, M is the gravitational mass of one body, m_g is the gravitational mass of another, r is the distance between them. To work out their acceleration you then need Newton's second law.

So we have now two definitions of mass, the gravitational mass (m_g) and the inertial mass (m_i) and it seems although they are

derived or found in very different ways (statically and dynamically)
they have the same value, which is why a falling body has a constant
acceleration irrespective of its mass. Combining Newton's gravity and
second law

$$F = \frac{GMm_g}{r^2} = m_i a \qquad (1.7)$$

If we assume that $m_i = m_g$ (the equivalence principle) then this
simplifies to

$$a = \frac{GM}{r^2} \qquad (1.8)$$

and so the acceleration depends only on the Earth's mass and not the
mass of the falling object. How Galileo would have loved to see this
formula! It is important in principle to see how things can be measured. Gravitational mass is a real thing because it can be measured.
You can use the equation above: drop any mass a known distance r
from a body mass M and the acceleration (and a knowledge of G)
will tell you the gravitational mass M. Note that this is dependent
upon our knowing G, so all that can really be done is measure relative
gravitational masses. Similarly for inertial mass. The way to measure
that is to know the force on an object and measure its acceleration.
Then use $F = m_i \times a$ to get to the inertial mass (m_i).

Newton's theory had two main problems. The first was that
a mechanism for gravity was not apparent, why should it follow
an inverse square law with distance? What physical process was
involved? Newton refused to speculate about this saying "Hypothesis non fingo" or "I do not feign hypotheses". He believed that if the
cause was not clearly determinable from the data then one should be
silent about it. The second problem was the pail experiment which is
the first real statement of the conceptual part of the problems that
have led to the idea of physics from the edge.

In the Principia, Newton described the following experiment. Fill
a bucket with water and suspend it from a ceiling with a rope, twist
the rope and then release it and let the pail spin. At first the pail
spins and the water does not because its inertia is hard to overcome,
and its surface is flat. Eventually though, the pail's spin is transferred

to the water by friction and the water spins, and its surface becomes concave: it bends up towards the edges. This occurs because of inertia, which, as Newton was the first to realise, makes it difficult for objects to change their speed and direction (they find it hard to accelerate, which includes spinning). This is why the vector quantity velocity is needed instead of the scalar quantity of speed. Objects like to travel in straight lines, the law of inertia, but the spinning water cannot because it is held within the walls of the bucket. The best the water can do is move outwards towards the wall of the bucket. So, the water piles up at this edge until a pressure gradient force due to the greater height of water forms that is large enough to push it back towards the centre and a stable concave water surface is the result.

Newton asked: what causes the curved surface of the water? It cannot be the rotation of the water relative to the walls of the pail, because at the start the water was rotating relative to the pail (the

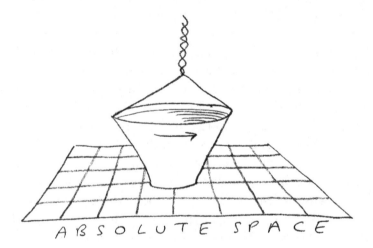

Figure 5.   **Newton's pail.** The bucket is spun, and the water stays still at first, then friction with the moving wall of the bucket makes it spin too. The water tries to move in a straight line since it has inertia, and so moves outwards but the walls of the bucket stop it and it piles up near the wall. How does the water measure its own movement? Relative to the walls of the bucket? No, its concavity was zero when the wall was moving and it was static. Its concavity is maximum when it is moving relative to its surroundings, and Newton said it was moving relative to absolute space — a bit like an imaginary grid.

pail was rotating relative to the water which is the same thing) and the concavity was zero. The concavity was largest when the relative rotations of pail and water was zero, they were rotating together (Hoffman, p40), but they were rotating relative to their surroundings. So Newton said that the concavity must be caused by the rotation of the water with respect to something he called absolute space. He imagined that space was like a stage upon which events occurred and could be rigidly defined using x, y, z co-ordinates.

Mach

B. G. Berkeley and G. Leibnitz hated Newton's concept of absolute space, but it was Ernest Mach who was the first to discredit it. He hated any concept that could not be directly observed, and he used Newton's own third law to attack the idea of absolute space. This law states that

> *For every action there is an equal and opposite reaction.*

I could never understand why this law was needed when I was at school, it seems both obvious and useless, but it has its uses, for example (Gleick, 2003, p133) if the Earth pulls on the Moon causing it to orbit, then the Moon also tugs on the Earth causing the Earth also to move, more slightly, and to show tides in its fluid components.

Mach argued that if absolute space was somehow producing a force that causes the inertia of the water, then surely there must be a reaction of the water back onto the nearby bit of absolute space, but by definition Newton had assumed that absolute space was unaffected by whatever went on within it, it was a background only, so this was a contradiction.

Instead Mach got around this problem by saying that the water was interacting with all the other **mass** in the universe, and especially the more distant matter since there was more of it. This gets around the problem because although the effect of all the other matter in the universe on the water is significant, the reaction acceleration caused by the tiny amount of water in a bucket on all the other matter in the universe is incredibly tiny since the inertial mass of the universe is so huge compared to the mass of the water in the bucket. This solution

Figure 6.   **Mach's pail.** Unlike Newton, Mach abhorred the idea of the inertial force coming from a rotation or acceleration with respect to an eternal absolute space because there was no way that space could cope with a reaction back from the bucket in clear violation of Newton's 3rd law. He suggested instead that inertia was a force that appeared when something rotated or accelerated with respect to all the other matter in the cosmos. This way there could be a reaction of the bucket on the cosmos, but it would be insignificant because of the huge mass of the cosmos relative to the pail.

does of course produce another problem and that is: how does the communication between the water and the rest of the distant universe happen, if special relativity forbids faster than light travel? We will come back to this problem later on, and fail to resolve it!

Mach also suggested a way to define inertial mass or experimentally determine it. He suggested that the best way to do this is to have two bodies side by side that attract gravitationally or magnetically (the origin of the force is immaterial, forgive the pun). They will then move towards each other, and the ratio of their accelerations will be the same as the ratio between their inertial masses (Bradley, 1971, p110). Mach also said that the principle of inertia is included

Figure 7.   Mach did not approve of Newton's concept of absolute space, which is a thing that can never be seen. He preferred to define theories like "we're moving" relative to observable things, like solid rocks.

in this since if there is no second mass, then there is no acceleration of the first, so its velocity is constant. A constant velocity is the default mode.

Before we leave Mach entirely, it is worth mentioning a fertile attitude to science that he had which influenced special relativity, and is also used later in this book. Mach stated that Newton's concepts of absolute space and time could not be defined in terms of observations you could ever make. Therefore they were meaningless. They were "purely a thought-thing which cannot be pointed to in experience" (Isaacson, 2007, p84). This 'don't trust things that can't be seen' attitude was very important to the early Einstein, but to discuss that we first need to talk about light.

Chapter 2

Modern Physics

Einstein

The greatest advances in physics in the 20th century are, for me, quantum mechanics and special relativity and not general relativity, for reasons that will be explained later.

To understand special relativity in all its strangeness we need to go back to the laws of motion we are familiar with and that were summarised by Galileo. It was proposed by Galileo that if A is a static object, (in our time) say a man waiting for a bus, and B is a bus moving at a velocity u, then the positions and the time of B (x', y', z', t') relative to that of A (x, y, z, t) are given by (see also Feynman, 1977, Chap. 15):

$$
\begin{aligned}
x' &= x - ut \\
y' &= y \\
z' &= z \\
t' &= t
\end{aligned}
\tag{2.1}
$$

These formulae agree with what we see happening to solid objects in the world around us. If we are driving on the motorway at 70 mph, and someone crawls past us at a relative speed of 30 mph, then using the first equation, in one hour they will be at a position $(x - ut)$ of $0 - 30$ mph $\times$ 1 hr $= 30$ miles ahead of us, and will probably have a police car after them.

We can draw this quite clearly as shown in Figure 1. The x axis represents the distance travelled and the y axis is the time. The car travelling at 70 mph is the left hand diagonal line, and the faster car is the line with the lesser slope. If we want to know how much further the faster car has travelled after 1 hour we just need to start

16   *Physics from the Edge: A New Cosmological Model for Inertia*

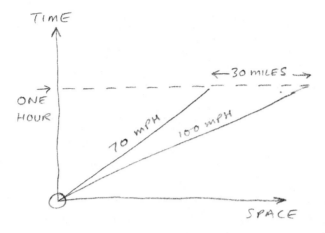

Figure 1.   **A spacetime plot.** Space is plotted on the x (horizontal) axis and time on the y (vertical) axis. A line on the plot represents the trajectory of an object.

the lines at the origin, watch them diverge for an hour and measure their distance apart on the horizontal axis: in this case 30 miles.

How does this apply to light? Although the ancient Greek Empedocles first claimed that light travels (Sarton, 1993) the influential Greek philosopher Aristotle disagreed and eventually the Greeks thought that light was emitted from the eyes, and since the distant stars could be seen light must have infinite speed.

Experimental proof that light travels at finite speed was first provided by Ole Romer in 1676. He realised that when the Earth is moving away from Jupiter in its orbit the time elapsed between the occultation of Jupiter's moon Io by Jupiter and its later emergence would be lengthened, since the light from the emergence of the moon would have further to travel than the light from the initial occultation (the time interval is shortened when the Earth is moving towards Jupiter). The amount by which the time interval is lengthened divided by the distance Earth has moved along its orbit is the speed of the light. Romer calculated the speed of light to be approximately 220,000 km/s. We now know it to be 300,000 km/s. Unfortunately, he was ignored at the time.

In 1873 James Clerk Maxwell wrote down his four equations that described how the electric and magnetic fields play off each other and he predicted that this interplay would lead to an electromagnetic wave with a speed of 300,000 km/s. To his amazement this was the speed of light. Thus it was found that light was an electromagnetic wave. But there was a paradox implicit in Maxwell's equations and Galilean relativity. In 1899 or so Einstein imagined what would happen if he followed a light beam at its own velocity. If you did that you would expect to see the electric and magnetic fields standing still in space, but he realised that Maxwell's theory does not seem to allow anything like that, the electromagnetic wave always has to travel at the speed of light!

Other people wondered whether light would follow Galileo's rules or not and Michelson and Morley in the USA attempted to measure the effect of the motion of the Earth on the light waves. By differentiating the above Galilean equations with respect to time, one would expect that the velocity of the Earth in its orbit (about 30,000 m/s) which is 0.01% of the speed of light (c) would mean that a light beam going along with the orbit would be 0.01% slower than one going across the orbital trajectory. In 1881 Michelson and Morley decided to test this using an interferometer as shown in Figure 2. Here, a light beam is split by a half-silvered mirror into two components, one into the aether wind (against the flow of Newton's absolute space) and one perpendicular to it. The light that travels upstream and downstream (along the Earth's orbit) can go in a straight line

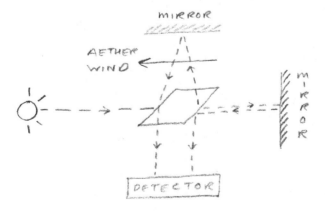

Figure 2.   Schematic showing the Michelson–Morley experiment.

and although it would be slowed up by the aether wind on its forward journey, it would be sped up by it on its way back, and these effects cancel. However, the light that goes perpendicular to the Earth's orbit (up and down on the schematic) was expected to travel further since it has to travel along a diagonal path to get back to the mirror, which has now moved (you can just about see these diagonals on the schematic). So it was expected that when the two beams of light got back to the half silvered mirror and were recombined, since they have travelled a different distance there and back, the phase of the individual waves in the light beam should not coincide. Instead, they always did coincide. Since the individual light waves are very short, about 10^{-6} metres, this method should have detected even the speed of the Earth in its orbit of 0.01% of c.

This was a shocking anomaly. Were the new Maxwell's equations wrong? This was unlikely since they seemed to agree very well with other observations. Maybe the older equations of Galileo were wrong, and there was a powerful taboo against this since people had long tested these equations so that they seemed to be almost common sense. This is why the solution found by Einstein (1905) was so shocking. He proposed that spacetime itself distorts so that the speed of light is always measured to be the same. More generally he said that:

All the laws of physics are the same in all inertial reference frames.

An inertial reference frame is one which is not accelerating, and the phrase "the laws of physics" include the law that the speed of light is 300,000 km/s. This means that the Galilean relativity laws described above, which are so ingrained in our experience of solid everyday objects, do not work for light and all electromagnetic radiation. There is no real surprise, since why should the mechanics of solid objects work for light, but this difference had escaped human notice up till then since light moves so fast that it takes centuries for societies to even notice that it moves at all (Adams, 1990).

The exact mathematical transformation required to modify space and time to keep Maxwell's equations the same for all observers was first suggested as a 'real' distortion of objects by the Irishman George FitzGerald and given mathematical precision by the Dutchman Hendrik Lorentz, prior to Einstein's more physical insight. Together they

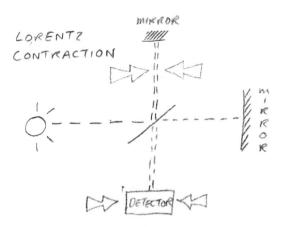

Figure 3. Lorentz contraction. In order that no change in the speed of light can be detected by the Michelson–Morley experiment, the metric of space itself must contract in the direction of movement. For light travelling at the speed of light the left-right contraction (see the schematic) is complete and this gets rid of the extra diagonal distance the light had to travel in going to and back from the upper mirror.

replaced the Galilean transformations mentioned above, with these transformations for an object moving at speed u in the x-direction:

$$x' = \frac{x - ut}{\sqrt{1 - u^2/c^2}}$$
$$y' = y$$
$$z' = z \tag{2.2}$$
$$t' = \frac{t - ux/c^2}{\sqrt{1 - u^2/c^2}}$$

where c is the speed of light. These are now called the Lorentz transformations. FitzGerald and Lorentz seemed to believe that the length changes were real changes in the objects relative to absolute space (a real shrinking), but Einstein, following Mach, disliked the aether, which was "purely a thought thing which can never be measured", and also was sceptical of the concepts of space and time, which also cannot be measured without the use of light. He proposed that instead of changing Maxwell's electromagnetic laws (which one can test in a laboratory) one should be elastic with the fundamental

20   *Physics from the Edge: A New Cosmological Model for Inertia*

Figure 4.   Lightning strikes a stationary train (picture above). The light gets to the person sitting in the centre of the carriage at the same time. They say it is simultaneous. If lightning strikes the ends of the carriage of a train moving forward the light from the lightning at the front of the carriage will get to the observer before the light from the strike at the back, so they will say the strikes are not simultaneous. Einstein was a follower of Mach who said that if something is not observable, it does not exist. This also means that whatever you can 'see' is the reality and there is no such thing as simultaneity, or that 'time is in the eye of the beholder'.

concepts of space and time which one cannot measure directly anyway. The meaning of the above equations is that space and time (x and t) change in order that c (which can be measured more directly) is kept constant for all observers.

For example, if lightning strikes the ends of a train at rest, see Figure 4, a passenger in the middle carriage would see each strike simultaneously, but if the train is moving forward the passenger would see the fore lightning strike first since he would be moving towards the light. So a word like "simultaneous" depends on your reference frame (how fast you are moving) and there is no absolute right answer to the question of "where did the lightning strike first?", it becomes relative (hence the name) to your own motion. Therefore just as Mach had argued, there is no absolute Newtonian time that

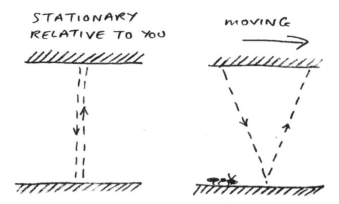

Figure 5.   **A mirror clock.** There are two mirrors, one above and one below (see the shaded areas) and a light beam bounces at the speed of light repeatedly between them. All observers must see the light travel at the speed of light c whether they are travelling with the mirrors or not (Einstein, 1905) so if the mirrors move, the frequency with which the light beam hits a mirror, a measure of time, has to slow down.

exists for everyone, each reference frame has its own time and the same goes for space.

One simpler way to think about special relativity is to imagine a clock made out of a couple of mirrors and a beam of light that bounces between them. One bounce means one tick of the clock.

The situation is shown in Figure 5. Two mirrors have their reflective surfaces pointing at each other and a beam of laser light bounces between them. Now, Einstein (1905) made the assumption, which has been successfully confirmed by experiment that, providing you are not accelerating, all the laws of physics must be unchanged for you. This is very similar to Galileo's idea that a person on a ship would not notice the motion of the ship, unless an acceleration occurs, which, on a ship, I have found, it always does (I went on a scientific cruise to Iceland in a stormy autumn once and was sick for most of the time). The laws that must remain constant include Maxwell's electromagnetic laws that force the speed of light to be $c = 300,000\,\mathrm{km/s}$. Einstein's assumption was provoked by his thought experiment about chasing a light beam, and the apparent observation by Michelson and Morley (1887) that the speed of light did not vary because of the Earth's motion through space, which he definitely knew about (Pais, 1982).

You can determine intervals of time because every time the light beam hits, say, the upper mirror the clock ticks and you can arrange things so that it ticks once per second. To produce a clock measuring seconds, since the speed of light is so fast, the mirrors would have to be 150,000 km apart. This is roughly the distance to the Moon, but we can do this in a thought experiment.

Now, imagine that suddenly this mirror clock is put onto a moving vehicle and driven past you (see the schematic on the right). Now we are forced to keep the assumption that the speed of light can't change, you have to see the beam of light moving at the same speed (c) because this can be measured and it seems it is always constant, but now (see Figure 5) the light has more distance to travel in the same second, since the clock is moving sideways the light is now travelling along a diagonal, so for someone looking at it, the time measured on the clock appears to slow down. The ticks occur less frequently so it appears that time on the clock has slowed down. Also, for a bug living on the clock, since his brain works using electromagnetic waves, his thought processes are also slowed down so that he does not feel that the clock is running slow at all. It all seems the same to him.

The main point here is that you might think that the slowness of the mirror clock is just an apparent thing because you are seeing it from afar, but this is not the case! This has been tested. Some scientists (Hafele and Keating, 1971) left one atomic clock at home and took one for a ride on a fast and high plane to slow it down and speed it up by special and general relativity respectively. When they brought the clocks back together the effects of relativity were still there: the clocks didn't agree. This is amazing and confirms the physical intuition of Einstein over the mere mathematical approach of Lorentz, because it means that the slowing down of time is "real" (whatever that means) and not just apparent.

This has a huge implication that goes back to my earlier mention of Berkeley and Mach: that reality is what you **can** observe. It seems that because it is impossible in our reference frame to ever perceive (using light) the moving mirror clock going at the 'normal' speed, then it doesn't go at the normal speed, it goes at the only speed we can perceive it to go, i.e.: slower. It is not particularly that we as humans cannot see it (it is not anthropomorphic or subjective in this way), but rather that this cannot be seen "in principle", a more objective view.

How does all this apply to mass? Einstein further showed that to make the change to Newton's laws all you have to do is change the inertial mass in the following way

$$m = \frac{m_0}{\sqrt{1 - v^2/c^2}} \qquad (2.3)$$

where the rest mass m_0 is the inertial mass of a body in its own rest frame (but in order to measure inertial mass you have to get it to move in response to a known force, so it is unclear to me what m_0 means). This formula looks innocuous enough, but it has a huge implication. Note that if you put $v = c$ into the formula the denominator becomes $1 - 1 = 0$ and the inertial mass (m) becomes infinite. This is the origin for the relativistic prediction that a spaceship, or any object, can never pass through the speed of light since its inertial mass becomes infinite and you need infinite energy to push it across the speed of light barrier. This does not mean that things that were travelling faster than light in the first place are disallowed, it just means that objects like us travelling slower than light, are not able to exceed it. This seems to preclude interstellar travel in a human lifetime, though as we shall see later there may be a way out here.

A speed of light limit is also puzzling cosmologically. When one thinks about it, it seems that the night sky should be bright. If we assume that the stars are uniformly bright and homogeneous (uniformly spread out) then the stars' brightness from where we are should decrease following a $1/r^2$ law, where r is the distance, but the number of stars at each distance away is proportional to the surface area of a shell around us which is $4\pi r^2$, so there are far more of them far away and this compensates for the fact that they are dimmer. What this means is that the night sky should be white, but it is dark.

The darkness of the night sky occurs, it is thought, because of the opaque gas and dust but also the Hubble expansion of the cosmos. The farther the star is away from us, the faster it is receding from us and the greater is the Doppler shift to its light. At some great

Figure 6.   A comment on the insight of Mach and Einstein that the world seems to depend to some extent on what we can see, using light, about it.

distance, the Hubble scale (which is thought to be about 2.7×10^{26} m) away, the stars are receding from us at the speed of light, their light is infinitely red shifted and we cannot see them. This is one way to account for the darkness of the night sky, but raises some questions.

One question is: what happens to a star at the Hubble edge? Just shy of the Hubble edge it is moving relative to us slower than light in the Hubble expansion, but a few minutes later beyond the Hubble edge it has accelerated and is now moving faster than light. According to special relativity, its inertial mass should have approached infinity at the speed of light somewhere in between these two times, so how did it manage to accelerate past this speed? In general relativity the reason given for this is that it is spacetime itself that is expanding, but this answer seems unsatisfactory to me since how can you test experimentally whether two stars are moving apart or spacetime is moving them apart? This seems to me to be a fix and I will come back to this problem later.

Figure 7.   **The Doppler shift.** An ambulance travelling faster than sound would not be immediately detectable using sound waves, just like stars at the edge of the cosmos are not detectable by light because they are moving away from us faster than light.

The Equivalence Principle

Special relativity only applies to unaccelerated motions which is not the most useful of limitations because this sort of environment only occurs in deep space. Einstein was unable to apply it to situations with gravity which is what we are used to on the Earth, but in 1909 he had what he described as his "happiest thought", and it was that if you are being accelerated at acceleration 'a' with no gravitating mass nearby, then you would feel exactly the same as if you were not being accelerated but close to a gravitational mass so that the acceleration it causes, $g = a$. So, if you are in an elevator that breaks free and falls down under gravity the two effects, of gravity and acceleration will cancel out exactly so that you float free, as people do on the International Space Station since they are free falling around the world. This is the same insight that Galileo had when he considered balls of different weights being dropped from the tower of Pisa.

As will be described in a minute, this led to the General Theory of Relativity which was a huge mathematical and perhaps also physical accomplishment. However, as will become clear later in this book, I regard this "happiest thought" of Einstein as being rather more based on aesthetical considerations than any real data. It is a beautiful idea, but there is nothing that forces the inertial mass and the gravitational mass to be equal, and I will spend the rest of the book trying to convince you that it isn't quite true, but that the difference between them is of such a kind that it has not been detected.

But, back to Einstein: imagine you are in a static elevator. You see a beam of light travel across in slow motion and it travels in a straight line. This is normal physics and Einstein says spacetime changes to keep the laws the same. Now imagine the elevator is accelerating upwards. Now a light beam travelling across your elevator would appear to have a trajectory curved downwards. Oh oh! The laws of physics seem to have changed, light is not moving in straight lines but is now moving on curves. So Einstein extended the principle of relativity so that the laws of physics must be unchanged in accelerated reference frames, but how do you unbend the trajectory of a light beam? Well, Einstein and Mach had no concern over bending space since it can't be seen anyway, so Einstein argued that space itself is curved (actually spacetime) and the light is moving on a straight line in bent space. If we do this the right way then the person in the elevator can never tell by measuring the bend in a light beam that they are accelerating since their rulers will distort too so as to hide the bend. This is what Einstein did with the General Theory of Relativity.

At the time (1915) there were other theories of gravity around. In 1914, for example, a Finnish scientist Gunnar Nordstrom, wrote down Maxwell's equations in five dimensions and then found that what came out was electromagnetism and gravity. He had thus unified these two processes easily. He got around the problem of the invisibility of the extra dimension by saying it was a circle so if you look along it you come back to where you started and the circle could be made to be too small to see.

However, Nordstrom and Einstein's theories were both tested by observations, and only Einstein's general relativity predicted the bending of light around the Sun and this bending was subsequently seen by Eddington in 1919, although the results at the time were probably not good enough to really prove this (Smolin, 2006, p39–43) subsequent measurements in high acceleration regimes have since also

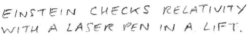

Figure 8.   Einstein's insight was that the laws of physics should be the same even in accelerated reference frames. This includes the law about the speed of light being the same and light travelling in a straight line. So, a laser beam in a lift should never allow us to measure its curvature. The curve of space would bend our rulers so we could never measure it.

supported general relativity (though not, I will argue, those in low acceleration regimes).

It is interesting that Nordstrom's approach of adding dimensions to try and unify theories managed to give a sensible looking answer, until it was falsified by the data, and this should be a warning to all, since it is possible to get the right answer for the wrong reason (Ptolemy's epicycles are another example) and only data can set the record straight. Unfortunately in my view Nordstrom's approach has been pursued by string theorists ever since and they are famously shy of any observation based tests. General relativity was also tested on the precession of the perihelion of Mercury and the bending of star light by the Sun. It has recently also been tested by experiments in orbit.

For example, the satellites used for GPS navigation, are travelling very fast, so that the onboard clocks should appear from the Earth to be going slower due to special relativity. The actual formula is

$$\frac{1}{\gamma} = \sqrt{1 - \frac{v^2}{c^2}} \tag{2.4}$$

which is the fractional amount of change due to special relativity for an initial value of 1 induced by moving at speed v. For GPS satellites which move at 3.9 km/s this change is a very small number: a 1 part in 10^{10} slowing, and this has been confirmed, and is routinely corrected for. Also, these satellites are in a medium Earth orbit 20,000 km high and very high up in the gravitational field so according to general relativity the time as measured by them should speed up by a factor given by

$$\frac{1}{\gamma} = \sqrt{1 - \frac{2GM}{rc^2}} \tag{2.5}$$

For GPS satellites at a height or radius (r) of 20,000 km general relativity causes a speed up of 5 parts in 10^{10} and this is also observed and is corrected for.

Gravity Probe B (GP-B) was launched in 2004 to measure two predictions of general relativity, frame-dragging and the geodetic effect (NASA, 2005). The GP-B satellite contained four gyroscopes cooled by liquid helium to 2 K to make their niobium coating superconducting. These gyroscopes were expected to maintain their orientation in space and each had a telescope on it looking at a guide star, so any change in their orientation could be seen.

The geodetic effect is a consequence of the distortion of spacetime by a large mass like the Earth and should cause a much larger effect on the gyroscopes. You can picture this by drawing a circle with the same radius as the Earth. Now put the Earth inside and according to general relativity space itself bends so that the circumference of the circle is reduced by 1.1 inches. Now because of this a gyroscope will change its orientation as it goes around (Stanford University, 2013). The expected change in their orientation due to general relativity was 0.0018 degrees per year. If Newton was right instead then the gyroscopes (not subject to any external force) would not change their orientation at all. The geodetic effect was confirmed by Gravity Probe B.

The second effect is the frame-dragging effect and is caused by the large mass of the Earth dragging spacetime around with it as it spins. It was expected to cause the orientation of the spin axis of the orbiting gyroscopes to follow the Earth's rotation by 0.000011 degrees per year. This effect has not been confirmed by GP-B.

General relativity has also been tested using binary pulsars, the first of which was discovered by Hulse and Taylor (1975). These very close pulsars are predicted to lose orbital energy by general relativity in the form of gravitational waves. These are waves in spacetime itself that propagate and take energy away from orbiting binaries. The observed decay of the orbit of Hulse and Taylor's binary was found to be consistent with general relativity (Taylor and Weisberg, 1982), but this is only indirect evidence for gravitational waves and general relativity. The waves themselves have not yet been seen despite the many detectors that have been built.

One of the problems I have with general relativity is that it still suffers from the problem described by Mach in that it depends on something that cannot in principle be observed: spacetime, and further when applied to the motion of galaxies and galaxy clusters it requires the introduction of 10 times as much mass as is seen to enable it to predict the rotation curves correctly. As will be discussed below this can be fixed by adding dark matter in a halo around the galaxy, but the particle required to give this diffuse halo has not been seen, and since it must be diffuse it cannot explain the similar behaviour of tiny globular clusters (see Chapter 3).

How can we do better? A clue may lie in the inconsistency between general relativity and quantum mechanics, especially since the latter is a far better tested theory and has agreed far more accurately with experiment. This is reminiscent of the clash between Galilean mechanics (taking the part of general relativity) and Maxwell's equations (taking the part of quantum mechanics) in which Maxwell's equations proved to be the more fundamental. Is quantum mechanics more fundamental? Well, it is certainly more bizarre...

30   *Physics from the Edge: A New Cosmological Model for Inertia*

Quantum Mechanics

When the temperature of a body is above absolute zero on the Kelvin scale, as all bodies are, the particles within it emit thermal radiation (all charged objects emit radiation when accelerated and temperature is just acceleration on small scales). Since the radiation is absorbed and emitted several times by atoms and electrons within the body, by the time it leaves the surface the spectrum is continuous and looks rather like the curves shown in Figure 9.

For a colder body the radiation emitted has a longer wavelength (the curve on the right) and less energy is emitted overall (the area under the curve is smaller). For a hotter body more energy is emitted (the area under the curve is larger) and the wavelength is shorter (the left hand curve).

The emission of heat radiation also depends on the character of the surface. If it is mirror like (as in a thermos flask) then both the emission and absorption of heat radiation is low and this is a general rule, a good absorber is a good emitter. A body which is a perfect emitter and absorber is called a black body because it would look black even if you shone a light on it (Ohanian, 1985, p902). One way to make a black body is to have a large hollow cube (a cavity)

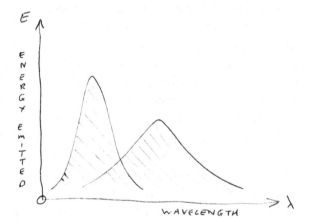

Figure 9.   **The energy content (vertical axis) of different wavelengths (horizontal axis) of radiation emitted from bodies above absolute zero.** The two peaks show the spectrum of radiation emitted by a hotter (left) and cooler (right) black body.

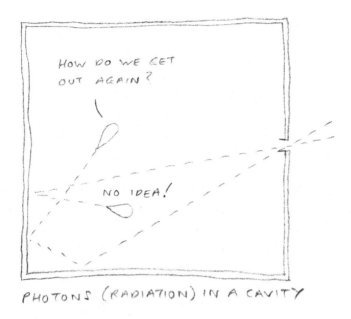

HOW DO WE GET
OUT AGAIN ?

NO IDEA!

PHOTONS (RADIATION) IN A CAVITY

Figure 10.   A black body cavity, like a maze, easy to get in, not easy to get out.

with a small hole in it. The hole is then a good analogy to a black
body because if radiation goes in through the hole, it bounces around
inside the large box and is unlikely to get out again, and if radiation
comes out, it stays out. See Figure 10.

Lord Rayleigh calculated the amount of thermal energy present
in such a cavity and found that an infinite number of waves were able
to fit into the cavity — they were limited because they all had to
have nodes at the boundary but there were still an infinite number of
them because you could have even very fine, short wavelength, high
energy waves. According to the equipartition theorem, each mode
or wave has an energy of kT (this is Boltzmann's constant k times
the temperature in kelvin) and so the energy for all of the waves
together was infinite. This was called the ultraviolet catastrophe,
which referred to the fact that most of the energy resided at the
short end of the wavelength spectrum, in the ultraviolet.

Planck (1901) solved this problem by proposing that energy is
always quantised. If an oscillator or wave has a frequency of ν then
the allowed values of energy are, $E = 0$, $h\nu$, $2h\nu$, $3h\nu$, etc., where

the h is now called Planck's constant. All other values for the energy are forbidden (waves that do not fit nicely within the boundary). This solves the problem because the thermal energy in the walls is shared between the different oscillators in the walls, but for the oscillators with very high frequency ν, the energy of the higher allowed energy (e.g.: $5h\nu$) is so large that even random variations of the energy in the walls won't provide enough energy to excite them, so the higher energies remain unexcited, and the ultraviolet catastrophe was averted (Ohanian, p905).

This looks very simple in hindsight but it was a revolutionary suggestion, not taken seriously until Einstein provided more direct evidence to support it five years later using the photoelectric effect.

From Planck's law it is possible to prove that the spectral emittance of a black body has a maximum at a wavelength given by

$$\lambda(\text{m}) = \frac{\beta hc}{kT} \sim \frac{0.0029}{T(\text{K})} \qquad (2.6)$$

This is called Wien's law, which was derived theoretically before it was tested experimentally, and in it appears a constant β which is known to be 0.2, Planck's constant h also appears and is $6.6 \times 10^{-34}\,\text{Js}^{-1}$, c is the speed of light, and Boltzmann's constant k is $1.38 \times 10^{-23}\,\text{JK}^{-1}$. This formula predicts the peaks of Figure 9 and says that if the temperature of an object (T) is low, then the waves of radiation coming off it are long. Conversely, if the temperature (T) is high, then the radiation coming off it is short.

For example, the Sun's surface temperature is $5778\,\text{K}$ so the radiation coming off it is shortwave radiation of wavelength $5 \times 10^{-7}\,\text{m}$. By not such a coincidence this is the wavelength of light we are used to seeing. We have developed through evolution to see the most common wavelength available. In contrast the Earth is a lot cooler, about $14°\text{C}$ ($287\,\text{K}$) so it emits longer wave radiation of wavelength 10^{-5} metres.

In 1905 Einstein took Planck's quanta more seriously and showed that they explained some curious observations made of the photoelectric effect. Lenard in 1902 showed that the energy of electrons ejected from a metal by light depended not on the intensity of the light, but its frequency. This supported Planck's quantum hypothesis since if you increase the intensity of light, it means you are firing

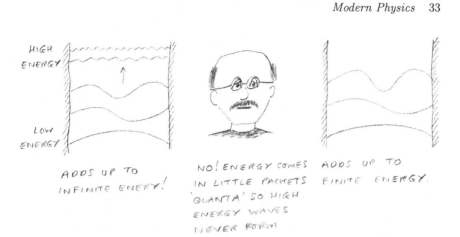

HIGH
ENERGY

LOW
ENERGY

ADDS UP TO
INFINITE ENERGY!

NO! ENERGY COMES
IN LITTLE PACKETS
'QUANTA' SO HIGH
ENERGY WAVES
NEVER FORM

ADDS UP TO
FINITE ENERGY.

Figure 11.   Planck's quantum idea. Allow waves of any wavelength (energy) and
you can build up infinite energy within a cavity (left). If you recognize that energy
comes in packets (the quantum waves must fit exactly within the walls), then it's
unlikely that a packet will have enough energy to excite the short (high energy)
waves and the problem disappears.

more light energy at the metal and can eject more electrons, but
the light energy comes in packets or 'quanta', so each electron can
only take one packet of energy so the electrons' energy is unaffected. If you increase the frequency of the light though, the packets
themselves can hold more energy and so the ejected electrons can
obtain more.

Planck had assumed that only the oscillators in the walls of the
cavity were quantised, but Einstein proposed that radiation itself
was quantised even when it was zooming around in the air and that
waves were composed of say $h\nu$ or $2h\nu$ amounts of energy. Planck
didn't like this very much, but the idea eventually prevailed and the
packets of energy were called photons. Einstein thought of radiation
as being rather like a gas of photons (Ohanian, p907).

This led eventually to the ideas of Louis de Broglie who proposed
that not only did radiation behave like a particle, but also particles behave like waves, so that when you fire, for example, protons
through two slits, even though they are supposed to be solid, they
interfere with each other and produce an interference pattern on a
screen beyond the slits. This is often called wave-particle duality and
shows up the apparent schizophrenia of the cosmos.

An Interlude About Hoyle's Cosmology

We have now looked at inertia, gravity and quantum mechanics. Gravity and quantum mechanics are modelled very differently in physics. Gravity is modelled using a smoothly curved spacetime, whereas quantum mechanics is modelled using the exchange of virtual bosons, particles. One is smooth, the other is pointlike. Which is more fundamental?

There is a problem with Newtonian mechanics and also general relativity in that if it were perfectly true then, surely by now all the matter in the universe should have self-attracted and collapsed into one huge mass. In contrast what we see is a universe that is not only not collapsing on itself, but it is actually accelerating away from itself!

In order to escape from a gravitational mass M, a body must be expanding away from it at more than the escape velocity (v) which is given by

$$v = \sqrt{\frac{2GM}{R}} \tag{2.7}$$

where G is Newton's gravitational constant, M is the mass of the attracting body and R is the distance to its centre. Putting numbers into this equation for the Earth, for example, you can calculate that the escape velocity is about 10.4 km/s, which is pretty fast, so rockets are needed to escape from the Earth's gravity.

It is possible to calculate the escape velocity of the visible universe, since we can count the number of stars we can see and make a crude estimate of its gravitational mass ($2 \times 10^{52\pm1}$ kg), and determine the distance to its edge (cosmic radius $= c/H = 1.3 \times 10^{26}$ m). Using the equation for the escape velocity above we get

$$v = \sqrt{\frac{2G \times 2 \times 10^{52}\,\text{kg}}{1.3 \times 10^{26}}} = 143{,}000\,\text{km/s} \tag{2.8}$$

Given the uncertainty in the values used for the mass and the Hubble scale, this velocity might be a low as 45,000 km/s or as big as 451,000 km/s so it is consistent with the speed of light of 300,000 km/s. This may be a coincidence, but it is also interesting. Why might the speed of light be an escape velocity for the cosmos?

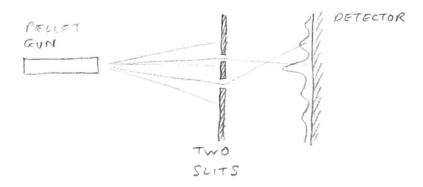

Figure 12.   The two-slit experiment works for solid particles as well as photons, but only for particles small enough that quantum mechanical effects become visible.

In some regard it does make sense since anything faster than light would presumably be able to achieve a universal escape velocity and would not be visible, hence the size of the observable universe.

It is possible to imagine lots of other combinations of mass M and radius R for which the escape velocity is greater than the speed of light and therefore the object would be black since any light falling on it would never escape to reveal its presence to us. The combination of the self-attractive nature of gravity and the eventual huge escape velocity is the idea behind a black hole.

Rearranging the above equation gives:

$$M = \frac{c^2 R}{2G} \tag{2.9}$$

This is the equation for the mass of the observable universe proposed by the Steady State theory of Hoyle (1948). The way I think about this theory is that it is based on the speed of light being constant, which is a property that has been observed, but this idea is extended to the speed at the universe's edge. If we assume that c must be constant there then as the radius R of the cosmos increases, then to keep the speed of light (the escape velocity) constant, the mass of the cosmos has to increase to keep the matter at a further distance bounded to the centre. So Hoyle's theory predicts that as the universe expands, it gains gravitational mass, following the equation above.

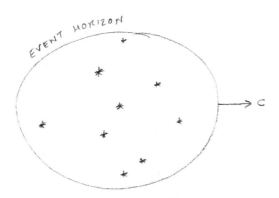

Figure 13.   **A very simple cosmology**, that is based on the idea that the speed of light is an escape velocity. This agrees with the idea that when something is moving away at the speed of light you cannot see it, so it has escaped from your knowledge of it. To make this work, as the universe expands you must increase the gravitational mass inside it to keep all the matter bounded to the centre and keep the speed (c) constant on the edge.

The theory was eventually discredited because it did not predict or explain the Cosmic Microwave Background (CMB) radiation, which appears to be a remnant of a time when the universe was extremely hot, but as we shall see later there may be a way to reconcile Hoyle's intuitive cosmology with the CMB.

Hawking Radiation

Generally the current theory of gravity, general relativity and quantum mechanics are thought to be incompatible, but there is one incidence where they have been combined to make a prediction: black holes. The theory of QED (Quantum Electro-Dynamics) predicts that virtual particles, including virtual electrons are constantly being created all over the place. The same is true at the boundary of a black hole. What Hawking (1974) noticed was that these electrons may then break up into two virtual photons at the edge of the black hole, going off in different directions, one entering the black hole and one zooming off into space. This separation caused by the event horizon makes the exiting photon suddenly real. He showed that the temperature of this radiation is

$$T = \frac{\hbar c^3}{2GMk} \qquad (2.10)$$

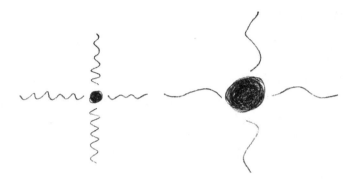

Figure 14.   **Two black holes.** One is small (on the left) and so is hot and radiates shortwave radiation (remember Wien's law) and the other one is massive, cold and radiates longwave radiation (on the right).

The derivation is given in Appendix B. This formula is odd because it says that a light black hole is hot, and a large one is cold. Therefore, small ones radiate heat to their surroundings and thereby lose energy and mass and get smaller, get hotter, and radiate more until they presumably vanish in a blaze of heat radiation. In contrast large ones are cold and, if they are colder than their surroundings they will not radiate at all. The surrounding temperature is the temperature of the Cosmic Microwave Background and is about 2.7 K, so we can do a quick calculation to see what mass the black hole has to have so that it does not radiate:

$$M = \frac{\hbar c^3}{2GkT} = \frac{1 \times 10^{-34} \times (3 \times 10^8)^3}{2 \times 6.6 \times 10^{-11} \times 1.38 \times 10^{-23} \times 2.7} = 5.4 \times 10^{23} \, \text{kg}$$

$$(2.11)$$

This is less than the mass of the Moon. It tells us that only black holes with masses less than that of the Moon will radiate and evaporate, bigger ones just absorb heat and energy and so get more and more massive (Adler, 2006). It will be argued later in this book that our universe is just such a black hole, growing and absorbing heat from its surroundings and giving rise to the acceleration of the cosmos: what has been called dark energy.

Figure 14 shows two black holes, one is small and so is hot and radiates shortwave radiation (remember Wien's law) and the other one is massive, cold and radiates longwave radiation.

Unruh Radiation

A black hole is not the only way to create an event horizon. Whenever
an object accelerates, a so-called Rindler event horizon appears. This
is because if you have an acceleration a, for example to the right,
information from a certain distance to your left has no hope whatever
of catching up with you, provided you continue to accelerate, since
the speed of information transfer is limited to the speed of light. The
distance away from you (r) of this Rindler horizon is given by

$$r = \frac{c^2}{a} \tag{2.12}$$

Fulling (1973), Davies (1975) and Unruh (1976) derived a similar
formula to Hawking's black hole horizon, for this dynamical event
horizon and found that the temperature (T) seen by an object with
acceleration a was

$$T \approx \frac{\hbar a}{2\pi c k} \tag{2.13}$$

A derivation is given in Appendix C. What this means is that, as
the acceleration increases so does the Unruh temperature and also
the wavelength of the Unruh radiation shortens. We can use Wien's
law that we looked at earlier, which tells us the wavelength emitted
from a black body of temperature T, and write down an expression
for the wavelength in terms of the acceleration:

$$\lambda = \frac{4\pi^2 \beta c^2}{a} \sim \frac{8c^2}{a} \tag{2.14}$$

This Unruh radiation usually has a wavelength too long to see with
our technology. For example, for a terrestrial object with an acceleration of 9.8 m/s^2 the Unruh wavelength is 7×10^{16} m. This is about
ten light years.

However, Unruh radiation might be visible for very large accelerations. Smolyaninov (2008) discussed an experiment by Beversluis
et al. (2003) who illuminated a gold nanotip with laser light and measured anomalous infrared photoluminescence coming off the nanotip.
He suggested that this was Unruh radiation. This is plausible since
the acceleration of the electrons around the tight curve of the tiny
nanotip is huge, it is $c^2/r = 9 \times 10^{22}$ m/s^2, where c is the speed of

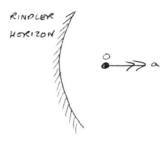

Figure 15.   An object accelerating to the right at an acceleration a, sees a Rindler horizon form to its left at a distance away of c^2/a.

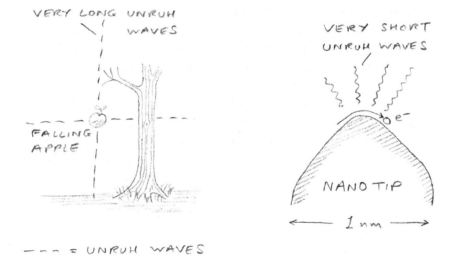

Figure 16.   Terrestrial accelerations, for example apples falling off trees, produce Unruh waves too long to be detected by our present technology, but electrons accelerating around gold nanotips, may produce Unruh waves short enough to be detected.

light and r is the radius of curvature and for such an acceleration the Unruh wavelength is $8c^2/a = 8 \times (3 \times 10^8)^2/(9 \times 10^{22}) = 8 \times 10^{-6}$ m, within the detectable infrared range. This is close to the wavelength that was seen.

The meaning of all this relatively recent work in physics for me is that whereas first we had a physics based on mechanics and thermodynamics, probably because people were used to thinking in terms of machines because of the industrial revolution, physics is now moving into an era where information is used, thanks in part to the changes in thinking led by the innovations of the computer age. This change in thinking may help to solve some of the problems that present machine-based physics cannot solve.

There are two ways of doing physics, one was pioneered by Isaac Newton and involves looking at data, or better still finding a 'crucial experiment' that focuses on the phenomena you want to understand and derive a theory from what you find. This is the empirical method. Then there is the method used by Einstein where you find a paradox such as following a light beam and noting that an electromagnetic wave cannot then exist. Paradoxes are great places to start thinking from because they represent the collisions, like in geology with continental plates, of two different paradigms of thinking, neither of which are true. At this interface, the deeper concept driving them may be seen, i.e.: magma.

I prefer the empirical way, but with a twist: the importance of anomalies. Theory can only get us so far and can be an obstacle to progress. The imagination of nature is far superior to ours so it is not enough to try to imagine the mind of God or try to use logic to derive new theories. I think it is necessary to go looking at real data to try and figure out what is going on, and to get to new physics this has to be anomalous data. That is where I shall go next.

Chapter 3

Problems at Low Acceleration

The Orbital Balance

We live on the Earth, and in this environment there are accelerations everywhere. Atoms are vibrating with heat, Brownian motion
is jostling molecules, sound and pressure gradients are vibrating and
moving the air, cars are zooming past and the Earth itself is spinning
and orbiting the Sun. A typical acceleration on the Earth's surface
depends on latitude. At the equator it is $v^2/r \sim 460^2/6300{,}000 \sim$
$0.03\,\text{m/s}^2$ and it decreases towards the poles. In free fall it is $9.8\,\text{m/s}^2$.

Progress in science often comes from looking at new regimes.
For example Galileo used the newly invented telescope to look at
Jupiter's moons for the first time and witnessed the, then unknown,
laws of orbital motion and he spent the rest of his life getting into
trouble for this.

Thanks to Kepler, Newton and the small correction of Einstein
we now understand orbital motion very well. When a smaller body
orbits a large one, they are attracted together by gravity, but if the
smaller body is orbiting, then its inertial mass is always trying to
make it travel in a straight line, and this inertial tendency pushes
it outwards from the larger body. This is what we call a centripetal
force. The orbits that we tend to see must be stable, otherwise they
would have collapsed, so there must be a balance between these two
opposing tendencies given by

$$\frac{GMm_g}{r^2} = \frac{m_i v^2}{r} \qquad (3.1)$$

where the left hand side is the force due to gravity pulling masses with
gravitational masses of M and m_g together over a distance r and the
right hand side is the inertial (centrifugal) force pulling them apart
and in this term I have written the mass as m_i. This is the inertial

42   *Physics from the Edge: A New Cosmological Model for Inertia*

Figure 1.   A small body (black circle) orbiting a larger. There is a balance between the gravitational force (G) pulling the smaller body inwards, and the inertial force (I) trying to get it to move in a straight line and pulling it outwards.

mass, which is usually considered to be equal to the gravitational mass m_g, but there is no reason why they should be equal.

If we do assume that the inertial and gravitational masses are equal then we can simplify the above equation to

$$v = \sqrt{\frac{GM}{r}} \qquad (3.2)$$

And this predicts the orbital velocity (v) of the smaller object m round a larger one, and it predicts that it should decrease in a particular way as the distance between them (r) increases. This formula has been extremely well tested on most of the planets in the Solar System. It predicts, for example that for it to remain in its present orbit, the orbital velocity of the Earth around the Sun must be

$$v = \sqrt{\frac{6.6 \times 10^{-11} \times 2 \times 10^{30}}{1.496 \times 10^{11}}} \sim 30\,\text{km/s} \qquad (3.3)$$

Which, luckily, it is. If it wasn't the Earth would not be in orbit. Assuming we have the mass and distances right, Jupiter's orbital velocity around the Sun is predicted to be

$$v = \sqrt{\frac{6.6 \times 10^{-11} \times 2 \times 10^{30}}{5 \times 1.496 \times 10^{11}}} \sim 13.3\,\text{km/s} \qquad (3.4)$$

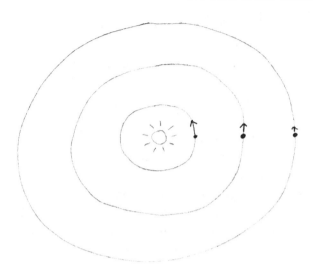

Figure 2.   The planets of the Solar System are locked in by a balance between the gravitational force towards the Sun, and the inertial force pulling them out. For this balance to work there must be a drop off of orbital speed (v) with distance from the Sun, otherwise inertial mass would win and the planets would zoom off into interstellar space. This can be seen in the plot, since the innermost planet has a large velocity (large arrow), and the outer planet has a smaller velocity.

which is rather slower. The point is that the orbital speed decreases as the orbital radius goes up, because as the gravitational attraction inwards lowers as the planet gets further from the Sun, the centrifugal force outwards required to balance it also can be smaller, so the orbital speed can be less. This formula, which relies on the assumption that inertial mass is equal to the gravitational mass works very well in the Solar System, although there is still some doubt as to the outer planets since some of them haven't completed a full orbit since we have been monitoring them.

Does this balance work on larger scales? There is no guarantee it will, and the first hint of trouble came from a Swiss astronomer called Fritz Zwicky.

Dark Matter

Zwicky had something of the character of Sheldon in the popular TV program "The Big Bang Theory" in that he was probably more brilliant than most of his colleagues, but he had a habit of telling them, even to the extent of carrying around with him a petition that he used to ask people to sign saying that he should be awarded the Nobel prize. He also used to call his colleagues "spherical bastards" since, he said, they seemed to have this property no matter which way he looked at them. Probably as a result of his abrasive character many of his discoveries were not readily taken up by others and had to be rediscovered many decades later and he never got his Nobel prize.

Zwicky had looked at the Coma clusters of galaxies. Galaxies tend to congregate in groups or clusters and there is a rule called the Virial theorem that states that the amount of kinetic energy present in a stable system is usually about half the gravitational potential energy, and this can be written:

$$2\,\mathrm{KE}_{\mathrm{avg}} = -\mathrm{GPE}_{\mathrm{avg}} \tag{3.5}$$

Astronomers are very keen to work out the mass of distant systems, and they can do this by looking at the light emitted, and assuming that the mass to light ratio is the same as that of the Sun. This is not a very direct way. What if the ratio of bright stellar masses to dark gas for example is different? Or if the stars are not Sun-like? This method would not work.

An alternative method is to use the Virial theorem, since the left hand side of the above equation, the kinetic energy, can be determined from the average speed of the stars (this is something that can be directly seen, so Mach would approve) and the right hand side is rather like the gravitational potential energy on the Earth, which is given by mgh, so you can work out the gravitational mass (m) from it (which you are trying to find). Zwicky looked at the speed of the galaxies in the Coma clusters and he noticed they had a huge amount of kinetic energy and he worked out the mass that this implied for a stable system, and it was 400 times larger than the mass implied by the stars he could see in the cluster — the light. The conclusion he came to was that there was a lot of dark matter in the cluster: a lot of matter that was gravitating, but not lit up (Zwicky, 1933).

Jan Oort in 1939 also noticed that the distribution of mass that he was able to infer from velocities in an elliptical galaxy didn't look like the distribution of light. This too was ignored.

Finally, in the 1960s and 1970s, Vera Rubin, managed to show more conclusively that the velocity distribution of the Andromeda galaxy showed a constant orbital velocity with radius. This is very different to what is predicted by the formulae used for the Solar System above which predicts a decrease of orbital velocity with increasing radius. What could cause a constant rotation curve?

Combining Newton's second law and his law of gravity but keeping the gravitational mass and inertial mass distinct, we can write

$$v = \sqrt{\frac{GMm_g}{rm_i}} \qquad (3.6)$$

Zwicky, Oort and Rubin saw that the velocities of galaxies and galaxy clusters were far too large at their edges. There are several ways to explain this.

The most popular solution and the one suggested by Zwicky is to increase the mass M in the formula above, by saying that there is simply more matter there than we can see. This seems reasonable enough (but reasonableness is not a proof that it is true). If more dark matter is in the galaxy then the outer stars can orbit far more quickly and despite their greater centrifugal outwards acceleration, they can still be bounded gravitationally to the system.

This kind of thinking also has a precedent. In the 19th century the orbit of Uranus showed an anomaly which lead Alexis Bouvard to propose, and Urbain Le Verrier to more accurately predict, that an unknown mass (a planet) was perturbing Uranus' orbit gravitationally. This planet (Neptune) was discovered close to the location where Le Verrier had predicted it would be.

The hypothesis of dark matter is very similar to this, but whereas in the case of Neptune, it required the addition of a small amount of normal matter in the plausible shape of a planet, galactic dark matter (because of the square root in the equation above) needs the addition of at least ten times as much matter as is observed in the Milky Way, in a new form of matter and with new physics to go with it to explain why it takes up a bizarre halo-like configuration around the edges of galaxies. New physics would have to be invented to explain why it doesn't converge on the galaxy and spin in the same pancake-like formation.

The following (repeated) story is not intended to disprove dark matter, that is very difficult: as a theory it is not falsifiable. Rather this story is intended to show the dangers inherent in this kind of thinking. There was a similar argument in Galileo's time. Aristotle had said heavier objects fall faster, but Galileo noticed in a hailstorm that big and small hailstones fell together and later on he showed, using balls and inclined planes, why this must be: it requires inertial and gravitational masses to be equal. His contemporaries, to play safe and support Aristotle, said: "Aha! The big hailstones must have fallen from a greater height". This could have been right, but the clue was in the *ad hoc* way they had to set up the initial height of the various sizes of hail: there was no reason for it. There is also no reason why dark matter should be in a diffuse halo around a galaxy.

There is another inconclusive, but philosophical objection to the idea of dark matter, in that it is not falsifiable and not predictive. Dark matter can be placed anywhere to make whatever model you want fit the galaxy rotation data. It could be used to support any theory at all. In effect it has an infinite number of adjustable parameters, so it could not either be said to be predictive. This means that if you were given a picture showing the light distribution of a galaxy, then using dark matter you would not be able to predict the rotation curve. What you have to do is assume that general relativity is correct, and work out the dark matter distribution that would enable general relativity to predict the rotation curve. You have not predicted the rotation though, you have only predicted the dark matter that supports general relativity, and you can't test this solution because you can't see the dark matter. This is certainly not good physics as we have ever known it. It is unpredictive and not falsifiable.

The only chance those proposing dark matter, i.e.: just about everyone, have to return within the realms of science, is to detect some. This is difficult because if it exists, it doesn't appear to feel the strong or weak forces or the electromagnetic force. If it did we would have seen or detected it by now. But if dark matter particles have inertial mass they should push things that they bump into. This is why huge tanks of liquid xenon, for example, have been set up as dark matter detectors underground. This liquid is very cold, so any recoil caused by a particle of dark matter, can be more easily seen above the background of thermal noise and the experiments are also placed deep down in mines to avoid collisions from energetic cosmic rays. There is one such experiment in the UK, 1100 metres deep down the Boulby mine in West Yorkshire and one in the US, in northern Minnesota at 700 metres deep (Brooks, 2009). Despite a couple of decades of looking in this way, they have found nothing. This does not of course disprove the existence of dark matter, and it does not mean that it has been a waste of time either because very often experiments designed to look for one thing often turn up another thing which is just as useful, and also a negative result is just as useful as a positive one. However, the lack of detected dark matter particles so far does mean that it is high time alternatives were suggested and looked at.

Looking back at Eq. (3.6), there are some other ways to increase the orbital velocity v. We could increase the gravitational constant G for larger distances like those on the scale of galaxies and this has been tried in the conformal gravity theory of Mannheim (1990). I have actually met Mannheim and I always remember him fondly because in 2006 I was proudly attending my first astronomical conference at the observatory in Edinburgh, a workshop on "Alternative Gravities". The Royal Observatory, for obvious reasons, is on top of a very large hill, and I was dragging my suitcase slowly up the hill to the meeting. Mannheim kindly stopped his car and gave me a lift up the hill, saying that he was very sympathetic to the plight of poor civil servants like myself (as I was then). So that night to return the favour I read all about his theory. I didn't understand much of it, but I gathered that gravity becomes repulsive at large scales.

Another solution to the galaxy rotation problem, is to decrease the inertial mass at low accelerations while leaving the gravitational mass alone. This is rather problematic because according to Einstein the inertial and gravitational masses are supposed to be the same, but inertia has always been a rather neglected concept in physics, and not well understood. In 1983 Mordehai Milgrom from Israel presented an empirical model that can do either of these things, i.e.: it modifies either G or inertial mass. It is called Modified Newtonian Dynamics, or MoND for short.

Figure 3.   The plight of alternative solutions to the galaxy rotation problem.

MoND

Newtonian dynamics, that is Newton's second law and his gravity laws were inspired when Newton looked at falling apples on the Earth and conceived the link between that and the orbit of the Moon. He devised a model that fit them both. The model that fit turned out to be a force of gravity that varied according to the inverse square of distance. The accelerations in these systems are relatively high, being $9.8\,\mathrm{m/s^2}$ on the Earth and $1.6\,\mathrm{m/s^2}$ at the distance of the Moon. At the edges of spiral galaxies like the Milky Way, the accelerations are ten orders of magnitude smaller, on the order of $10^{-10}\,\mathrm{m/s^2}$, so to model these systems we are extrapolating Newton's laws down to accelerations that are equivalent to the acceleration of Pluto due to the gravity of Mercury (Scarpa *et al.*, 2006). This is a big assumption and the last time such an extrapolation was made over such a range of scales was when classical physics which holds at scales of 1 metre or so was applied to the atom which has a scale of 10^{-10} metres, and that didn't work at all and the very different quantum mechanics had to be developed (Unzicker, 2008).

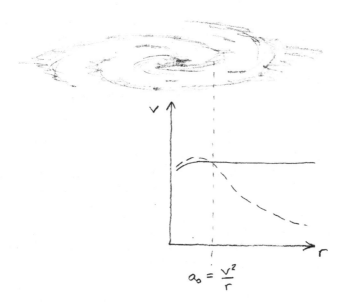

$$a_0 = \frac{v^2}{r}$$

Figure 4.   **The rotation curve of a typical galaxy.** The expected Newtonian drop off of orbital velocity is shown by the dashed line in the graph. The solid line is the observed velocity which does not drop off. The acceleration at which the two lines diverge is always $a_0 = 2 \times 10^{-10}$ m/s^2. This is difficult to explain using dark matter.

Milgrom (1983) noticed a peculiar pattern in the aberrant behaviour of galaxy rotation curves which is shown in Figure 4. The graph shows the Newtonian velocity (dashed line) which predicts a drop off of velocity with radius. The observed velocity (as shown first by Rubin) doesn't drop off and the divergence of the two curves, the deviation from Newtonian gravity, was always observed when and only when the gravitational acceleration computed considering only baryons (visible matter) fell below a value close to 2×10^{-10} m/s^2 which is called a_0 (Scarpa *et al.*, 2006). Making use of this observation Milgrom (1983) developed an empirical model that has two versions, the normal Newtonian part for high accelerations and a new one for accelerations below a_0. For accelerations that are below the crucial acceleration of 2×10^{-10} m/s^2 either Newton's second law is modified to reduce the inertia, or Newton's gravity law is modified to increase gravity.

For reasons that are based on observation, I will focus on the possibility that the inertial law is the one that needs to be changed. Milgrom did not specify which property should be changed, but did work out the changes necessary for the inertial possibility. At familiar accelerations there is the equation that everyone knows: $F = ma$. For accelerations below a_0 Milgrom suggested the following

$$F = m_g \mu(a/a_0)a \qquad (3.7)$$

where m_g is the gravitational mass, μ is a function that when a is large is equal to 1, so we get $F = ma$ back, and when the acceleration is below a_0 ($a \ll a_0$) the function is a/a_0 so that we get the new force formula of: $F = ma^2/a_0$. Here a_0 is the very small acceleration mentioned above. What happens in between these two functions, how to fit them together for intermediate values of a, is a matter of some debate, but this doesn't affect the results much.

Isaac Asimov once said that the most significant words in science are not "Eureka!" which means "I have it!", but rather: "That's funny...". No, I do not mean to suggest for a moment that MoND or Modified Newtonian Dynamics is funny, but rather that there is an interesting observed pattern here that the deviation from expected behaviour always starts at the acceleration a_0. The value of MoND is not that it explain this, it doesn't, but that it pointed out this pattern, that is crying out for an explanation.

The value for a_0 has a number of intriguing qualities about it. For example, it is close to the speed of light times the Hubble constant which is the fraction by which the universe is expanding (i.e.: $a_0 = 2 \times 10^{-10}\,\mathrm{m/s^2}$, and $cH_0 \sim 5 \times 10^{-10}\,\mathrm{m/s^2}$). The critical acceleration a_0 has the property, mentioned by Milgrom, that if we start with a particle at rest at the supposed beginning of the universe, 13 billion years ago, and gave it this acceleration, then by now it would have a speed close to that of the speed of light. This is a tremendously suggestive proposal, suggesting a link between dynamics and cosmology. The acceleration a_0 is also the simplest acceleration one can write down using the cosmological parameters, c and the Hubble scale Θ: c^2/Θ (Smolin, 2006). All these coincidences, and they could of course be coincidences, give one the impression that MoND is saying something very fundamental that is as yet too big to comprehend.

The problem with thinking about all this is that all the parameters are linked together. For example: are we so sure about the age of the universe? Of course not. No one, except perhaps Marvin the Paranoid Android (Adams, 1979), has been standing around for 13 billion years with a stop watch. If not, how can we be sure that a_0 gives you the speed of light in the lifetime of the universe? If you adjust one of the cosmological parameters, the others will have to be adjusted too and also they are all rather indirect measurements (Mach would put them in the category of being 'thought things' and not directly measurable). My feeling when I first read about it was that MoND is telling us something, but all the cosmological parameters are uncertain and maybe all of them are wrong, but the linkages between them that are implied by MoND fit a better model that was waiting to be discovered.

Perhaps a good idea would be, in the spirit of Mach, to come back towards concepts that can be directly observed (i.e.: direct observations) and link them to just one of the cosmological parameters. Then we are not assuming what the model that binds them all together cosmologically is (because we are not confident that we know what it is yet). So how about the observed galaxy rotation curves which can be seen and the Hubble scale Θ, which can be fairly directly deduced?

It was already clear from the data that MoND only seems to apply to the outer edges of galaxies. When the orbital acceleration of stars goes below a_0, then the dynamics become non-Newtonian. This is difficult to explain using dark matter because it means you have to put all the dark matter into the galaxy model as a diffuse halo around the galaxy. How can this be justified? Dark matter does not feel any force apart from gravity, so how does it stay diffusely spread out over such a huge halo? You have to invent a whole new physics for it, that enables it to remain spread out. But, of course, this is not proof that dark matter doesn't exist, just that it requires new physics also and this violates the principle of Occam's razor that the simplest solution is the best.

I then read one paper that convinced me, that the solution to galaxy rotation was not dark matter. This was by Scarpa *et al.*, (2006) and there have been other more comprehensive and conclusive ones published since, and ignored by astrophysicists, but this was the one I happened to read first. They looked at globular clusters which are small areas within the Milky Way, where the stars are arranged slightly more densely than in surrounding areas. Since dark matter is seen to be so diffuse, it should not affect the dynamics of small sub-galactic collections of stars like globular clusters. You can't have it both ways and argue that dark matter has to stay spread out on huge scales to fit galaxy rotations and then say it exists in small lumps within globular clusters. The dark matter should be uniform at these scales so should not affect dynamics. But, Scarpa *et al.* (2006) found exactly the same thing happened with these globular clusters as with larger galaxies: whenever their internal accelerations dropped below a_0, their dynamics became non-Newtonian. I still have this glorious paper and I wrote on my copy: "Brilliant stuff: tells me which way to go!". Also, the fact that it is just the internal accelerations that matter for dynamics, and not the external gravity field, shows that it is not MoND that is at work either, but that is for later.

If the change in behaviour is not due to dark matter, it is also difficult to explain using MoNDian variations in G, since why should G increase for low accelerations? Of course, it would be possible to invent a model for this, but there is a possible physical reason that can explain, in some sense, a change of inertia at a_0. Here is a quote from Milgrom (2005):

> *Can we spot an inkling of MoND inertia in the Unruh effect? When the acceleration of a constant a [acceleration] observer becomes smaller than a_0 ... the Unruh wavelength becomes larger than the Hubble distance. We expect then some break in the response of the vacuum when we cross the a_0 barrier...*

I was very taken with this idea at the time, and I called it "Milgrom's break" since the Hubble scale is 'breaking' the process of inertia. We can do the calculation very simply. We discussed the temperature of Unruh radiation (Eq. (2.13)). We also discussed Wien's law in the section on quantum mechanics (Eq. (2.6)). Putting these two together gives

$$a = \frac{4\beta\pi^2 c^2}{\lambda} \qquad (3.8)$$

Figure 5.   As the acceleration of an object (say, a rocket ship) decreases, the wavelength of the Unruh waves it sees increases. Milgrom proposed that when they exceed the Hubble scale (the size of the observable universe) they might vanish and this can be understood using the ideas of Mach who said that if something cannot be seen in principle, then it does not exist.

Now we can do the calculation. Putting in the size of the observable universe which is 2.7×10^{26} m, we get an acceleration of 20×10^{-10} m/s^2. Now admittedly this is ten times the value needed for MoND, but it is interestingly close.

This was where Milgrom left the problem. He did not specify why there should be a break in the Unruh radiation specifically, and did not manage to derive MoND from this idea or test it against observations. The idea of a cutoff is also not able to explain galactic dynamics since it implies a sudden change in inertia at a particular radius, not the slower change that is seen in the data.

If we go back to Mach, he said that if something could not be observed, then it cannot exist, and it is possible to justify Milgrom's break by this kind of argument. If an Unruh wave is larger than the Hubble distance, then it cannot be observed in principle and therefore cannot exist, and since inertia is related to the acceleration which in turn depends on the Unruh wave, then inertia also disappears. Put like that it sounds rather batty, but I like the logic of this: the idea sounds right, and someday thinking more in terms of information we may understand the context better.

Anyway, for now, no matter how suggestive Milgrom's break is, it does not exactly fit the observations. If it was true then as soon as a star moved far enough out in a galaxy, it would drop below the acceleration of a_0, its inertia would dissipate completely, and inertialess, it would zoom off into the cosmos at infinite speed. This is not what is seen: the change in galactic dynamics is more gradual than that and the edges of galaxies are filled with stars that, although they are behaving strangely, are still bounded to the galaxy and have some inertial mass. So a theory is needed that produces a more gradual decrease in inertia. Instead of Milgrom's *break*, we will consider a slight *bend*.

MoND is Not Good Enough

Whether or not it is Unruh-related, MoND itself, as an empirical model has some problems. The main problem with MoND is that it is not a theory. It is an empirical model that has been determined, rather like Planck's black body formula, by fitting a formula to galaxy rotation data with an adjustable constant a_0. This is not satisfying and an explanation needs to be given for the value of a_0 and why, if not a coincidence, it looks very much like cH_0 or c^2/Θ.

MoND also has problems with globular clusters. As I hinted above, these globular clusters, which are embedded within the Milky Way galaxy, show non-Newtonian behaviour when and only when the *internal* acceleration falls below about $1.8 \pm 0.4 \times 10^{-10}\,\mathrm{m/s^2}$ (Scarpa *et al.*, 2006). According to MoND it is the *total* acceleration (including the external acceleration) that is important, and these clusters are embedded in parts of the Milky Way where the total acceleration due to the galaxy is greater than a_0, and yet these clusters still show anomalous behaviour when their *internal* acceleration only goes below a_0. This does not agree with MoND at all.

If internal accelerations only are the key then it is good news in a sense because it means we can set up an experiment for which internal accelerations are less than a_0, even if the external accelerations due to the gravity of the Sun or the accelerations due to passing cars are much larger, and still detect the anomalous behaviour. Such experiments have already been done, and in some of them anomalous

Figure 6.   It is possible for a system such as a globular cluster, or a couple of people in a roller coaster to have a huge total acceleration (relative to something external), but still have a tiny internal acceleration. These two friends' acceleration relative to the Earth is causing one of them concern, but they are hardly accelerating relative to each other. For globular clusters, it is the internal acceleration that appears to matter.

behaviour has been seen, but they have proven difficult to reproduce in other labs so these result are interesting but, rightly, controversial. I will discuss these results later on and I do pay more attention to them than most scientists because there is a tendency in modern physics to discount experiments that disagree with accepted theory, and in science, it should be the other way around. Anyway, all this implies that we need a theory for which the *local* accelerations are used to determine inertia.

The first systems in which apparent dark matter was detected were galaxy clusters, which are gravitationally bounded villages of galaxies on a huge scale. Zwicky found that huge amounts of dark matter (400 times the visible matter) were needed to keep these galaxy clusters bounded (now we know it is more like 100 times).

MoND does not perform well on these clusters. If it is used then, still, dark matter, has to be added as well and typically 2 or 3 times the amount of visible matter needs to be added. So it seems that even MoND needs dark matter to save it at the largest scales.

So in summary, MoND has pointed out a very interesting pattern, but it is empirical and no reason was given for it, and it does not work for globular clusters, because their behaviour depends on their internal accelerations and it does not work for galaxy clusters. Something more fundamental is needed.

Chapter 4

A Solution from the Edge: MiHsC

The Physics of the Flute

I have played the flute since I was 12 and I am very fond of the instrument. Apart from the pleasant sound which is uniquely pure when you play it, since the sound waves are close to being pure sine waves, you feel connected to it, since to produce a sound you have to vary the air pressure you apply until you get a resonance from the instrument. It becomes a sort of acoustic interaction between the instrument and the player. This is unlike the piano for example, on which you push keys without so much feedback. As a result, one could play the piano without the ability to hear, and this is what Beethoven did after he became deaf. I suspect you could not do the same so easily with a flute though you could rely on vibration transmitted through the body.

The interesting thing is that the model for inertia I am going to talk about is very similar to the physics of the flute, which may be where I got it from. Anyway, I can use the flute to introduce the idea.

A flute is a long cylinder, completely open at one end, and also open at the other via a mouth hole along its side. The instrument is played by blowing over this mouth hole. Sometimes the flow of air goes over the hole and sometime it goes obliquely in, eventually an oscillation between these two patterns occurs, a wave on the air jet, and if the pressure wave from a particular incursion of air into the flute is able to go at the speed of sound to the other open end of the flute and back just in time for the next incursion, then the waves reinforce each other. This is called a resonance and it means that the waves grow, and this is how the flute produces systematic vibrations (sound) in the air that emanate from the open end.

58   *Physics from the Edge: A New Cosmological Model for Inertia*

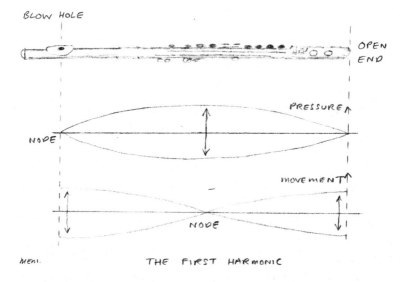

Figure 1.   (Upper picture) The flute is open at one end and partially open at the other. (Middle) the variation of pressure within the flute for the first harmonic: the longest wave. There is less variation in pressure at the ends since the air is connected to the atmosphere, but there can be variation inside. (Bottom) The movement of the air is at a maximum at the ends where the air is less constrained by the cylindrical walls.

If you want to produce a higher pitch then the waves on the jet must be closer together because the waves of sound in the instrument are closer together, their wavelength is shorter. So you must blow the air more quickly across the side hole, so that the waves pass across this hole at just the right frequency to match the sound waves inside the flute.

The pressure at the open ends of the flute must be atmospheric pressure and cannot vary by much, but inside the flute they can vary. The simplest arrangement like this is shown in Figure 1.

Figure 1 (upper picture) is a drawing of my flute. The curve just below that represents the variation in air pressure at various points along it for the longest wave possible in it. The pressure varies most in the centre and least at the open ends, since at the ends air is always available from surrounding areas to fill in low pressure, and higher pressure air can diverge into the open space. These points of minimum variation are called nodes. The curve below that shows the movement of the air. This is maximum at the open ends since there are no walls there to damp the motion, and least in the middle.

The wave shown in the picture is the longest air pressure wave that is possible in the flute and this means it is the lowest note possible, a middle C. This has always caused me occasional disappointment since many nice tunes (depending on the key they are written in) have been written with the B below middle C in them, so when I try to play music not specifically written for the flute, I often have to skip a note. This is a nice analogy to the long wavelength cutoff of Milgrom discussed above. Just as the flute does not allow notes with a wavelength greater than its length, the observable universe does not allow Unruh waves greater than its size, but the flute analogy allows us to go a bit further than that.

If we draw either the pressure wave or the displacement wave at an instant in time it would look like Figure 1. So the fundamental, the lowest note has a wavelength that is twice the length of the flute. This is because in order to match the next wave that appears at the blow hole, the wave has to travel from one end to the other and then back again. Instruments like pan pipes, which have a specific length, play just the fundamental and other waves that fit into the cyclinder. Pan pipes are arranged so that they have cylinders of different length so the players can chose the length of the cylinder and the note they want to play. Trombonists can vary the length of their cylinder to vary the note. On the flute you can vary the note in two ways. The first is to blow harder, and by this you can excite the second harmonic or the third, etc. This would move you up an octave from middle C to top C and then the G above that. These harmonics are shown in Figure 2.

These are the longest waves you can have so that there is a pressure node at the open ends of the flute. The longest is middle C, the next one is the C above that. The next one is actually a G, then high C then an E and so on. This is called a harmonic series and if the frequency of the first note is f, then the frequencies of the others are: $2f$, $3f$, $4f$, $5f$, etc.

So how do you play a note between middle C and upper C? There is a very nice trick to this. What you do on the flute is open a hole at a point along the flute and so enforce a pressure node at that point.

60   *Physics from the Edge: A New Cosmological Model for Inertia*

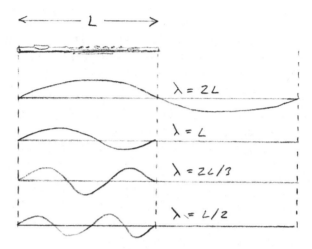

Figure 2.   **Pressure waves in a flute.** From top to bottom, the first, second, third and fourth harmonics.

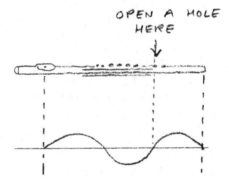

Figure 3.   How to force a particular note on the flute by opening a hole.

Only, preferably, one wave can satisfy this node, so you can select notes (Wolfe, 2012) as shown in Figure 3.

To summarise, only sound waves with wavelengths shorter than twice its length are possible in the flute, just as in Milgrom's break. However, in the flute the conditions are even more stringent in that only waves that fit *exactly* into the flute (or twice the length) are allowed, and all others are disallowed. Opening holes along the length of the flute makes the conditions even more stringent and further waves are disallowed. This necessity for an exact fit of the wave within the system, is the basis of the new model of inertia presented in this book. This is a one-dimensional example, but I will now discuss a closer two-dimensional analogy involving parallel plates and also ships at sea.

The Casimir Effect

Heisenberg's uncertainty principle can be written as $\Delta E \times \Delta t = \hbar$ which means that the uncertainty in the energy (ΔE) times the uncertainty in time (Δt) equals a constant which is Planck constant over 2π ($\hbar$). So if we consider very short times (small Δt) then a larger amount of energy (ΔE) is available for borrowing for that short time. As a result, if you look at a vacuum in which there is apparently no mass-energy, you won't see a vacuum since the vacuum can "borrow" energy for a short time to produce what are called virtual particles. There is another uncertainty relation: $\Delta p \times \Delta x = \hbar$ and this says a similar thing: the smaller the scale you look at (small Δx), the more uncertain momentum is (larger Δp) so particles with momentum appear out of nowhere on very small scales. This is called the zero point field, because although the energy is supposed to be classically at the zero point, there are still particles (also a field) caused by this quantum uncertainty, like waves on a sea. After de Broglie, we know that every particle can also be thought of as a wave (a field), just as the sound in a flute is both a wave and a note (different physics, but the same maths). If we take the electromagnetic field then how can we limit the waves as in the flute? The answer is to put two flat conducting plates close to each other and parallel. This means that the electromagnetic field must be zero at the two plates since the electrons in the metal move around to cancel it, just as the sound wave in the flute must be zero (have a node) at the two open ends as air moves to equalise pressure there. Just as in the flute only certain wavelengths are allowed so there are fewer photons between the plates than outside them.

This means that more photons bang into the plates from outside than bang into them from inside and since photons have momentum there is a net force pushing the plates together. This is called the

62   *Physics from the Edge: A New Cosmological Model for Inertia*

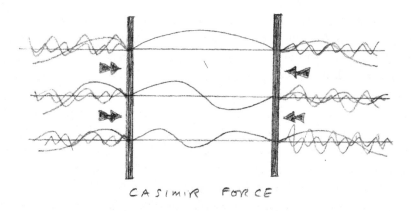

CASIMIR FORCE

Figure 4.   Two parallel conducting plates suppress zero point field wavelengths that do not exactly fit within the plates. Wavelengths outside the plates are unaffected, so there is a greater zero point field outside the plates and since the particles of this field have momentum, more of them bang into the plates from outside than from inside, pushing them together. This Casimir force was predicted by Casimir (1948) and observed by Lamoreaux (1997). It has been predicted that there is a similar force due to ocean waves that pushes two close parallel ships together at sea (see Boersma, 1996), but this is extremely small. It has been measured in the laboratory.

Casimir effect and the Casimir force after Hendrik Casimir who first predicted it (Casimir, 1948). It was observed just as predicted by Lamoreaux (1997).

An Asymmetric Casimir Effect (aCe)

It may seem strange that the first part of the new model I am going to talk about was proposed in the last paper I published on it (McCulloch, 2013), but conceptually this part comes first. Where does inertia come from? What is the mechanism by which Unruh radiation (part of the zero point field) causes inertia?

A model for inertia was proposed by Haisch *et al.* (1994). They proposed that oscillating partons (particles) within an accelerated object feel a magnetic Lorentz force, due to their interaction with the zero point field, that opposes the acceleration of the object. The force they derived was

$$F = \frac{-\Gamma w_c^2 \hbar a}{2\pi c^2} \tag{4.1}$$

Where Γ is the Abraham–Lorentz damping constant of the parton being oscillated, w_c is the Compton scale of the parton below which the oscillations of the zero point field have no effect on it, $\hbar$ is the reduced Planck's constant, a is the acceleration and c is the speed of light. Their derived formula does look rather like inertia, since the force is always in the opposite direction to the acceleration (there is a minus sign). However, their derivation was complex and it has been criticised on relativistic and other more basic grounds by, among others, Unruh, and Levin (2009). The main problem I have with this model is that the Lorentz force they used to generate inertia acts on the particles when they are oscillating at very high frequencies and since the frequency can go all the way up to infinity, they had a similar problem to the ultraviolet catastrophe we discussed earlier and they had to introduce an arbitrary frequency cutoff to get rid of the high frequencies, and that seems unphysical and is rather like a tuning parameter: never a good sign.

I will here propose a simpler model. As discussed above, the zero point field can be regarded as something like the ocean with waves and particles appearing all the time, in pairs to conserve momentum. Hawking (1974) looked at the production of these twin particles near black holes and found that the event horizon surrounding a black hole caused by its intense gravity was able to separate some of these pairs of particles. He thus predicted Hawking radiation and black hole evaporation.

It was then proposed by Fulling (1973), Davies (1975) and Unruh (1976) that a similar dynamic event horizon forms near accelerated objects. Consider a spacecraft that is being accelerated in the x-direction. At a certain distance opposite to the acceleration vector of the spacecraft, there is a surface beyond which no information can ever catch up with the spacecraft, the volume behind this surface can never be observed by the spacecraft (unless the spacecraft turns around of course!). This surface is called a Rindler horizon.

Just like a black hole event horizon, and just as in that case, it was proposed that virtual particles and radiation can be split up by this information boundary and realised into the real world. The result is that an accelerated body sees thermal black body radiation (a warm background) surrounding it, whereas an unaccelerated object

64   *Physics from the Edge: A New Cosmological Model for Inertia*

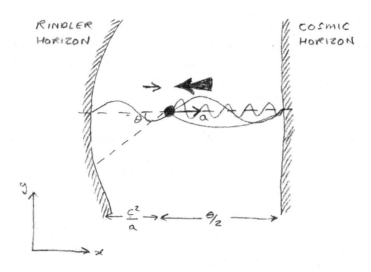

Figure 5.   A schematic showing a particle accelerating rightwards (the black circle). The shading shows the Rindler horizon close by to its left (at a distance c^2/a away) and the cosmic horizon far away to its right. The Unruh radiation is damped more on the left side producing an asymmetric Casimir effect (the force indicated by the two unequal opposing arrows) that pushes the particle (O) to the left against its accelerations: a model for inertia.

in the same space sees no radiation at all and concludes that the background is cold (the heat is caused by the Unruh effect).

Now imagine a particle being accelerated to the right, in the positive x-direction (see the schematic in Figure 5). An isotropic (from all directions) Unruh radiation appears all around it and radiation can exert a pressure. This is the principle behind solar sailing where photons from the Sun push on a thin solar sail. Now on the opposite side of the particle to the acceleration vector there is formed a Rindler horizon whose distance away is $d = c^2/a$, where c is the speed of light and a is the acceleration of the particle. Figure 5 shows a particle that is being accelerated to the right. A Rindler horizon appears on its left side. If we calculate the energy density of the Unruh radiation, in the direction of the acceleration, to the right, most of the Unruh waves will be allowed since the only event horizon there is the cosmic one far away at the Hubble distance (2.7×10^{26} m), but on the opposite side, to the left, fewer waves in the Unruh spectrum will be allowed because the dynamic (Rindler) event horizon is much closer, at a distance of c^2/a. For an acceleration of $9.8\,\text{m/s}^2$ the distance to the Rindler horizon would be $10^{15}\,\text{m}$, or about a tenth of a light year. So the momentum/impact of the Unruh radiation will be lower from the left (with a strongly damping Rindler-scale Casimir effect) and greater from the right. And this will push the object back against the applied acceleration. This asymmetric Casimir effect (aCe) models inertia intuitively.

This looks like inertia, but do the numbers work? The derivation is in McCulloch (2013) and summarized in Appendix D and the derived equation for the inertial force (dF_x) is as follows

$$dF_x = -\frac{\pi^2 ha}{48 c l_P} \tag{4.2}$$

(when I first published this formula I made a factor of two error, see McCulloch, 2013, that was kindly pointed out to me by Dr Jaume Giné, from Spain). In the formula h is Planck's constant, a is the acceleration, c is the speed of light and l_P is the size of the particle. If we put in the Planck length for the size, the inertial mass is predicted to be $2.75 \times 10^{-8}\,\text{kg}$: about 26% larger than the Planck mass.

This derivation is only valid for single particles. To calculate the inertial mass of a compound object it would be necessary to multiply the above mass by the number of particles present. It is also only valid for velocities much slower than that of light, since otherwise Lorentz contraction would alter the dimensions of the particle so that it would look more like a pancake.

This formula for inertia is considerably simpler, in both derivation and final form, than the formula of Haisch *et al.* (1994) and the mechanism is different. In the article of Haisch *et al.* (1994) the inertial process was a magnetic Lorentz force acting on particles oscillating at very high frequencies, and to avoid an infinite energy they had to impose an arbitrary upper frequency limit. The asymmetric Casimir effect proposed here produces an energy *difference* between the two sides of the particle, so no cutoff is needed.

How can this model be tested? Well, an entire chapter of the book will be devoted to potential tests, so I won't go into that very deeply here, but I have suggested (McCulloch, 2008b, 2013) that it may be possible to modify inertia by creating an event horizon using the metamaterials proposed by Pendry *et al.* (2006) and Leonhardt

(2006). They have shown that radiation of a given wavelength can be bent around an object (which must be smaller than the wavelength) using a metamaterial, making that object invisible to an observer at that wavelength. It may be possible instead, to set up a metamaterial to reflect radiation so that an artificial event horizon is formed. Then according to the model discussed here, this will damp Unruh radiation on one side of the object which would then be accelerated towards the horizon. It may be simpler to use metamaterials to directly damp the Unruh radiation on one side of the object.

Modified Inertia by a Hubble-Scale Casimir Effect (MiHsC)

So an asymmetric Casimir effect can explain standard inertia quite well, but it would be no fun if we stopped there since we need to predict something that is different from the standard model to start to test the new ideas. There is another horizon I haven't considered yet in any detail and that is the Hubble-scale horizon. Let us see what happens when we add this to the mix.

Figure 6 (from McCulloch, 2007) shows the energy spectrum of Unruh radiation seen by an accelerated object on the y-axis and the wavelength of that radiation on the x-axis. On the Earth, where accelerations are high, or on a highly-accelerated spacecraft, we will see short Unruh waves like those on the left and the background perceived by the object will be warm. The previous section showed that standard inertia is caused by the energy in the Unruh radiation spectrum (the area under the curve). Now, we will consider the effect of the Hubble horizon. The waves that fit into this horizon, like the waves that fit into the flute or the Casimir plates, are shown in Figure 6 by the vertical dashed lines and their wavelengths are $\lambda = 2\Theta/n$, where $n = 1, 2, 3$, etc. On the left of the plot the allowed waves are close together so most of the waves for high accelerations are allowed, the Unruh spectrum remains much as before and the inertia is hardly affected.

In contrast, for a body with a low acceleration the Unruh spectrum will look like the one on the right, and will lie on a part of the plot where there are fewer allowed wavelengths. In the plot the spectrum is only sampled by one allowed wavelength (2Θ) so only a

The Pioneer anomaly as modified inertia

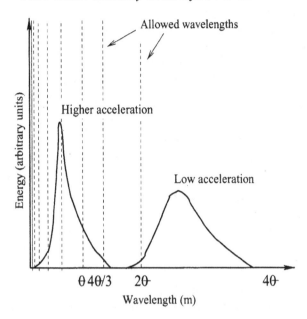

Figure 6.   How the inertial mass caused by the asymmetric Casimir effect (area under the curves) is modified, and filtered by the Hubble-scale Casimir effect.

very small proportion of the Unruh spectrum will remain and we can expect the inertia to decrease in a new way.

Why should only waves that fit exactly into twice the Hubble scale be allowed? I originally assumed it is due to a cosmic Casimir effect, but recently I have come to think of it philosophically. Knowledge from outside the Hubble horizon cannot be allowed in, so any pattern that is bigger than the boundary, or doesn't fit exactly within it, would give observers inside a clue as to what lies outside by extrapolation, and this cannot be allowed. I call a horizon like this a hard-horizon, since not only does it not allow information to pass through by movement, it also modifies waves that pass through it to make sure they don't allow information through either via their pattern. This more philosophical approach is mathematically equivalent to a Hubble-scale Casimir effect and, in my view, can be seen directly in the low-l Cosmic Microwave Background anomaly (McCulloch, 2014). I will explain this in more detail later.

Here we will assume that the unmodified inertial mass is just the area (A) under the curves as shown in Figure 6: $m_i = A$. How do we take account of the loss of energy in the disallowed wavelengths? Well, at first I took a guess and assumed it would decay linearly with wavelength, then I got scared realising that it would not be exactly linear, and calculated it with a Fortran program (a reviewer also pestered me to do this!). I tried varying the peak wavelength and counted the number of allowed wavelengths remaining in the subsampled curve where the spectral energy was more than 1% of the peak energy. It turned out that it was roughly linear after all. Anyway, we can add a factor to the formula to account for this linear decay of energy as it is subsampled less and less efficiently at longer wavelengths, like this

$$m_i \propto A \times F \tag{4.3}$$

What is the function F? Well, we can assume it looks like $F = (\lambda/\alpha) + \beta$ where α and β are two constants to be found. We know that when the wavelength is zero, then F must be one because the curve is perfectly sampled at short Unruh wavelengths and there is no energy loss, so putting this assumption that when $\lambda = 0$ for $F = 1$ into the equation we get: $1 = 0/\alpha + \beta$. So $\beta = 1$! We also know that when the wavelength is twice the Hubble diameter (4Θ) then F must be zero since we are just about to exceed the length of the longest Unruh wave and this is Milgrom's break. The only way the above equation works with this fact ($0 = 4\Theta/\alpha + 1$) is if $\alpha = -4\Theta$. So we have now

$$m_i \propto A \times \left(1 - \frac{\lambda_m}{4\Theta}\right) \tag{4.4}$$

The bracket is the correction we need to apply to the unmodified inertial mass, which is the rest of the right hand side so we have a new equation for inertial mass

$$m_I = m_i \left(1 - \frac{\lambda_m}{4\Theta}\right) \tag{4.5}$$

I should mention at this point briefly that this is a rough approximation. Actually, the number of allowed wavelengths will oscillate up and down in ever slower intervals as we increase the wavelength as the allowed wavelengths resonate with the boundary, but this will do for now. The λ_m is the peak wavelength in the Unruh radiation spectrum which is given by Wien's law (Eq. (2.6)) and if we put this value for λ into the equation for inertial mass we get

$$m_I = m_g \left(1 - \frac{\beta \pi^2 c^2}{a\Theta} \right) \qquad (4.6)$$

We can simplify this a little, introducing an error of about 1%, if we write

$$m_I = m_g \left(1 - \frac{2c^2}{a\Theta} \right) \qquad (4.7)$$

This is the formula we will be using for most of the rest of this book and it describes how the standard inertia from an asymmetric Casimir effect (McCulloch, 2013) is modified by a Hubble-scale Casimir effect (McCulloch, 2007). The complete model can be called Modified inertia by a Hubble-scale Casimir effect (MiHsC) or simply quantised inertia.

The m_I from MiHsC behaves in a similar way to what would be expected from MoND. Figure 7 is a graph showing the acceleration along the x-axis from tiny accelerations on the left to normal accelerations on the right and the ratio between the inertial and gravitational mass along the y-axis. For the assumption of an equivalence principle there is the straight dashed line $m_i = m_g$.

The prediction of MoND is shown by the dotted line and agrees with the equivalence principle until the acceleration is as low as fitting parameter a_0 (2×10^{-10} m/s^2) and then inertial mass reduces.

The predicted ratio from MiHsC is shown by the long dashed line and it approximates the equivalence principle for high (normal) accelerations, but reduces the inertial mass in a new way for tiny accelerations such as those at the edges of galaxies. The MiHsC curve is smoother than that of MOND and doesn't need a fitting parameter.

I should clarify that the acceleration in the MiHsC formula, is that with respect to other masses as suggested by Mach who said that only things that are observable should be put into a theory. For example, for an object on the Earth, its acceleration with respect to the Earth is zero, since it is attached. However, the Earth is spinning, so the object has an acceleration relative to everything else in the universe 'the fixed stars' which is latitude-dependent. I will prove the utility of assuming this later on.

70   *Physics from the Edge: A New Cosmological Model for Inertia*

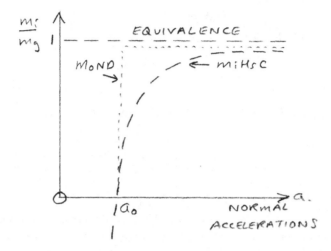

Figure 7.   A graph showing the acceleration along the x-axis and the ratio between the inertial and gravitational mass along the y-axis. For the assumption of an equivalence principle there is the straight dashed line $m_i = m_g$. MoND agrees with this until the acceleration is as low as a_0 and inertial mass suddenly reduces. MiHsC approximates the equivalence principle for high (normal) accelerations, but reduces the inertial mass in a new gradual way for tiny accelerations.

For an object in mid-latitudes the acceleration with respect to the fixed stars is about $0.03 \, \text{m/s}^2$. Given this acceleration, the inertial mass would decrease below the gravitational mass (by MiHsC) by

$$m_I = m_i \left(1 - \frac{2 \times 300000000^2}{0.03 \times 2.6 \times 10^{26}}\right) = m_i(1 - 2.3 \times 10^{-8}) \quad (4.8)$$

This a loss of 2.3×10^{-8} kg for every kg of gravitational mass: a very small effect, but experiments have been done to test the equivalence principle down to relative accelerations of $10^{-15} \, \text{m/s}^2$, so it is important that before we progress I explain why these experiments do not see the effect of MiHsC.

These experiments are usually done with a torsion balance. They measure the differential attraction of two balls on a cross bar suspended on a wire, towards distant masses like the Sun, by detecting tiny twists in the wire (see Gundlach *et al.*, 2007). This is done by rotating the torsion balance so the changing angle of the Sun is apparent in the result. This is like Galileo's famous experiment in which he dropped two balls from a tower. The heavier ball is more attracted towards the Earth (it has more gravitational mass), but also finds it harder to accelerate (it has more inertial mass). This applies to the balls in the torsion balance in their 'fall' towards distant masses. These experiments have indeed shown no twist in the wire, so you might say that in order to make such a claim as I do for MiHsC I would need some balls. However, the change in inertial mass predicted by MiHsC is independent of mass, as you can see by putting Eq. (4.7) into Newton's second law and his gravity law like this

$$F = m_i a = \frac{GMm_g}{r^2} = am_g \left(1 - \frac{2c^2}{a\Theta}\right) \tag{4.9}$$

which simplifies to

$$a = \frac{GM}{r^2} + \frac{2c^2}{\Theta} \tag{4.10}$$

Imagine Galileo dropping his two balls of different weight off the tower. MiHsC predicts that the change in inertial mass for both balls is in the same proportion at all times, in the above equation the extra MiHsC acceleration is independent of the mass, so they will still drop together, although MiHsC predicts that they will be accelerated by the Earth's gravity and drop ever so slightly faster than expected (the new term on the right hand side). Similarly, in the torsion balance experiments there will be no twist in the wire, as observed.

Figure 8.   With a nod to Douglas Adams, a poor sperm whale and a bowl of petunias debate the futility of knowing that MiHsC increases their acceleration very slightly.

Chapter 5

Evidence for MiHsC

In this chapter I will present comparisons of the predictions of MiHsC with quite a few observed anomalies, as tests. Now some of these anomalies are fairly solid, like cosmic acceleration and the galaxy rotation problem and some are not so solid. I am not saying by discussing them, that I necessarily believe that all the anomalies here are true ones, some may be errors, but I am saying that taken together they do seem to be pointing to new physics that looks rather like MiHsC.

Note that for most of these predictions it will be assumed that the standard part of inertia (modelled by the asymmetric Casimir effect in Chapter 4) is unchanged, and the Hubble-scale Casimir effect produces a slight deviation.

Cosmic Acceleration

The first prediction that can be gained from MiHsC can be obtained just as above where this expression was derived:

$$a = \frac{GM}{r^2} + \frac{2c^2}{\Theta} \tag{5.1}$$

Note that the standard Newtonian result includes just the first term on the right hand side, whereas MiHsC predicts the existence of a new term that depends only on the speed of light and the Hubble diameter. This term is tiny and is about $6.7 \times 10^{-10}\,\mathrm{m/s^2}$. This acceleration would take you from zero to $1\,\mathrm{m/s}$ in 317 years, from zero to 60 miles per hour in 8500 years, and intriguingly from zero to the speed of light in the age of the universe (13 billion years), so it is very small. However, this extra acceleration is important because it does not depend on the mass of the object or the mass of the object gravitationally attracting it. It stands alone. This means that when an object, for example, leaves the Solar System, and travels into deep space where there are no masses to exert a gravitational pull, there must still be this residual acceleration and also, as we shall see later, when special relativity enhances the mass of objects close to the speed of light, this acceleration is unaffected.

Where does the extra term come from? What happens is that as the acceleration of an object decreases as it, for example, moves out of the Solar System, the Unruh waves it sees and which make up its inertial mass, get longer, so that according to MiHsC a greater proportion of them do not fit within the Hubble-scale and so are disallowed by the Hubble-scale Casimir effect (see Eq. (4.7)) so the inertial mass drops and the object is then more easily accelerated by the same external gravitational force (the gravitational mass is unchanged). At some point a balance is reached between the ever increasing rate of loss of inertia due to MiHsC, and the acceleration due to constant external forces caused by that loss of inertia. This balance point due to MiHsC predicts that there is a minimum acceleration in nature of $2c^2/\Theta$.

Taking this to an extreme you can see that MiHsC does not allow zero acceleration because then the Unruh wavelengths would certainly be longer than the cosmos and unobservable (so disallowed,

Figure 1.   MiHsC implies that there is a minimum acceleration in nature and so disagrees very slightly with Newton's first law. This deviation is so small that it is very difficult to detect, and, unlike the figure would not even impact the slow world of snails. Nevertheless, in deep space, far from any forces, it can become important.

if we adopt the attitude of Mach, which I do). Therefore the inertial mass would be zero and accelerations would be infinite. So in MiHsC, as the acceleration approaches zero, inertia disappears slowly and infinite speeds are avoided.

As we will discuss later, the Pioneer craft have left the Solar System and are showing just this anomalous acceleration, but before we can be encouraged too much by this it is important to point out that the anomalous acceleration of the Pioneers can also be explained by an anisotropic radiation of excess heat from the spacecraft (Turyshev *et al.*, 2012). Although the analysis of Turyshev *et al.* might be correct, it is certainly not conclusive because their model is too complex to be published in full, has two adjustable parameters and contains over 2000 finite elements. MiHsC, in contrast, predicts the correct result without any adjustable fitting parameters and from a simple philosophy.

Do we have any other examples of masses moving in far deep space? Yes, at the edge of the universe there are plenty of objects called supernovae which are thermonuclear explosions of stars (Fraser, ed., 2006, *The New Physics*) and they are brighter, emit more light, than the galaxies they explode in so they are visible to us at great distances, hence their use as cosmic markers. Perlmutter (1999) and Riess (1998) found that supernovae very far away (at high redshifts) are fainter than one would expect if gravity was slowing the universal expansion down over time. If this gravitational slowing was occurring the supernovae would still be bright because they wouldn't have moved as far away from us. What could possibly offset the gravitational self attraction of the universe? Well, the standard answer to this is that it is 'dark energy' which is speeding the universal expansion up, and whose origin is unknown, but as we have seen MiHsC predicts this acceleration without needing any dark energy.

So, how does this work in the framework of MiHsC? I will not assume the Big Bang in this discussion because in my opinion it is better to stay focused more on things that can be as directly seen as possible. What I will do is consider a model of the universe as it is today, but put in more energy than is currently seen in the form of random motions and accelerations. Slowly this excess energy is

converted to heat and the objects in the universe slow down and
their accelerations reduce towards zero. This means that the Unruh
waves seen by all the masses in the universe get slowly longer and
longer.

However, in MiHsC, zero acceleration is not allowed because the
Unruh waves would then be larger than the size of the universe (the
Hubble scale) and therefore cannot exist (following Mach's principle)
so the longer Unruh waves disappear in a cloud of logic, the inertial
mass dissipates and it becomes easier to accelerate all the objects in
the universe and so they accelerate again. As discussed above, MiHsC
predicts that a balance is achieved around the universal acceleration
of $2c^2/\Theta$ and we eventually find that the acceleration of the cosmos
cannot fall below this value. If it did the Unruh waves would exceed
the size of the cosmos, inertia would disappear and the acceleration
would increase again. In effect you could argue that the energy for
dark energy is coming from a loss of Unruh-energy caused by the
Hubble horizon.

In my opinion this is a very physical, self-consistent and elegant
way to understand the existence of the cosmic acceleration and there
is a way to test it. The new second term in the equation depends on
the size of Θ and in cosmic history, since the universe is expanding
this scale has increased in time, so the cosmic acceleration must have
been much greater in the past.

This may help to explain the so called flatness problem which
was pointed out by Dicke and Peebles (1979). The problem is that
the observed mass of the cosmos is close to the critical value that
determines whether the universe is 'closed' and will eventually self-
collapse or 'open' and will continue to expand. In other words it is
'flat'. Guth (1981) proposed that the universe was hugely inflated
early on so that now it is many times larger than the parts we can
see and this explains its apparent flatness. MiHsC indeed predicts
that the acceleration must have been much greater in the past, but
I have not done the detailed calculations for this yet and also I tend to
mistrust cosmological extrapolations like this, and prefer tests closer
to home.

The second, MiHsC, term may also depend on a variable speed of
light c, but the evidence is too weak to decide on that at the moment.

The Large-Scale CMB Anomaly

In 1964 Arno Penzias and Robert Wilson discovered background radiation coming from all directions in space and this is now called the Cosmic Microwave Background (CMB) since its wavelength is in the microwave (a few micrometres). This radiation is just what you would expect from a universe with an average temperature of 2.725 ± 0.002 K, but it is not quite uniform. There is an asymmetry that is thought to be associated with the movement of the local group of galaxies with respect to this radiation (a Doppler shift), and when this is subtracted out there are smaller variations as well which are about a few parts in 100,000 of the background temperature.

Data from the Cosmic Background Explorer satellite (COBE), the later WMAP (Wilkinson Microwave Anisotropy Probe) and most recently the Planck satellite (Planck Collaboration, 2013) show a very close agreement with the Cold Dark Matter model (ΛCDM) at most scales, and particularly the small scales those with multipoles (a parameter given by the symbol L) greater than 40. Note that Λ represents dark energy, another invisible component of the model. It was shown by Hinshaw *et al.* (1996) for COBE and Spergel *et al.* (2003) for WMAP that there is an unexpected lack of angular two-point correlation on scales between 60 and 170° in all wavebands. This means that on large scales the CMB radiation is unexpectedly uniform, but these results were not significant. The recent results of the Planck Collaboration (2013) are very similar to those of WMAP (they agree within 3%) and the Planck team have again found that for multipoles (L) of: L less than 40 the CMB energy is between 5 and 10% lower than expected according to the best fit Planck ΛCDM model, and this anomaly is significant to the 2.5 to 3 sigma level.

It turns out that the Hubble-scale Casimir effect used in MiHsC to suppress long Unruh waves also predicts a suppression of large scale variations of the same size as seen by Planck. According to MiHsC the energy in the CMB should be modified as follows

$$E' = E \left(1 - \frac{\lambda_m}{4\Theta} \right) \tag{5.2}$$

where λ_m is the peak wavelength of the radiation and Θ is the Hubble diameter. The CMB data is usually presented using the concept of the multipole. The zeroth multipole, or $L = 0$, is a constant value across the sky, which would represent a wave much bigger than is allowed in MiHsC. The first multipole or $L = 1$ represents the longest wavelength that can fit into the sky, or the cosmos, this has a wavelength of 2Θ. The multipole moment can be written in terms of the Hubble diameter as follows

$$L = \frac{2\Theta}{\lambda_m} \qquad (5.3)$$

So that the correction from MiHsC is

$$E' = E\left(1 - \frac{1}{2L}\right) \qquad (5.4)$$

So, as usual with MiHsC, for higher multipoles or smaller cosmic scales there is no effect since the second term in this formula is negligible, but for low multipoles or large scales there is an effect: MiHsC suppresses variations on a large scale. The result is shown on Figure 2 as the dashed line.

The solid line in Figure 2 shows the observed angular two-point correlation of the CMB radiation as a function of the multipole moment (denoted L) as observed by the Planck Collaboration (2013). There is a large decrease in the power of the spectrum for $L = 2$, and 3, but this is not significant given the large errors caused by a lack of data at these large scales. What is significant, to 2.5 to 3 standard deviations, but not apparent on the graph, is the 5 to 10% reduction in power below that expected from the ΛCDM for all the multipoles (L) which are less than 40.

The dotted line shows the predicted spectrum using the best-fit ΛCDM model. This model fits the CMB spectrum very well for smaller scales ($L > 40$), but it overestimates the observed energy at the larger scales, where $L < 40$, by 5–10%. The data from Planck has shown that this difference is significant at the 2.5 to 3 sigma level. This means there is only a 0.27% chance it is due to chance and not an error in the model, so it probably represents a problem with the model.

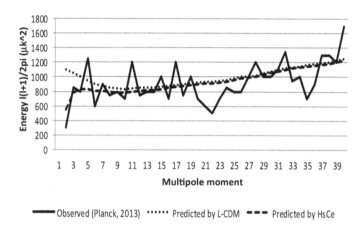

Figure 2. The observed spectrum of the CMB (the solid line) for multipoles below 40 as derived from the recent paper by the Planck Collaboration (2013). Also shown is the prediction of the Planck ΛCDM best fit model (the dotted line). In the report by the Planck Collaboration (2013) they concluded that there is: "*a tension between the best-fit ΛCDM model and the low-L spectrum in the form of a power deficit of 5–10% at L < 40, significant at 2.5–3 sigma. We do not elaborate further on its cosmological implications, but note that this is our most puzzling finding in an otherwise remarkably consistent dataset*" (Planck Collaboration, 2013). The dashed line shows the prediction of a Hubble-scale Casimir effect, which agrees with the observations.

The dashed line shows the prediction from the best fit ΛCDM model, but with a correction applied using the Hubble-scale Casimir effect using Eq. (5.4). This correction has no effect on the smaller scales where $L > 40$, but it does reduce the energy from the ΛCDM model for the largest scales ($L < 40$) by 5.5%. The modified prediction now agrees with the Planck observations which showed a deficit relative to ΛCDM of between 5 to 10%.

The meaning of this suppression is subtle: it is not simply a suppression of the longest wavelengths of the CMB radiation, since such wavelengths are far too long to be measured by our instruments, but is rather a suppression of the large scale variation of the radiation over the cosmos.

As briefly mentioned above, there is a way to understand this that uses and illustrates the philosophy I have used in MiHsC. The Hubble horizon is a horizon beyond which we cannot see anything in

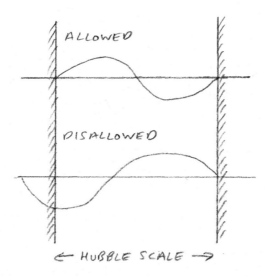

Figure 3. **Horizon censorship for the CMB anomaly.** Only patterns that fit exactly within the Hubble scale (with nodes at the boundaries) can be allowed. If all patterns were allowed then they would enable us to know what lies behind the Hubble horizon and this would defeat the purpose of the horizon.

principle (see Figure 3). Now if there is a pattern of variation that fits exactly within this boundary (Figure 3, see the upper line on the schematic) then this is allowed because the pattern is closed at the boundary and it doesn't imply anything beyond it. However, a pattern that does not fit exactly within the Hubble scale (see the lower line on the schematic) allows us to guess what lies beyond the Hubble boundary, and this should not be allowed. So what I am suggesting is a kind of horizon censorship. The horizon makes sure that it remains a boundary for information, even by reaching into the cosmos and disallowing patterns that might allow us to guess what lies behind it.

As a test of this idea, the Hubble-scale Casimir effect predicts that the spectral power should vary in a particular way for very low multipoles. The energy should show peaks when the waves fit exactly within the Hubble diameter, and troughs when they do not. It may be possible to search for these resonances in the Planck data if it is available at higher spectral resolution.

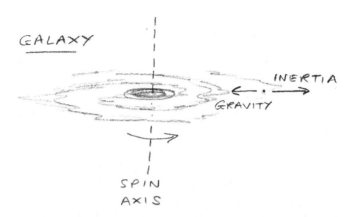

Figure 4.   **A typical disc galaxy.** The figure shows the forces on a star near its edge. The gravitational force (dependent on the gravitational mass) pulls the star towards the galactic centre, and the inertial force (dependent on the inertial mass) tends to keep it going in a straight line and therefore outwards from the galactic centre.

Galaxy Rotation without Dark Matter

The galaxy rotation problem is the best known anomaly in physics. To put it simply: galaxies should explode, but don't. That is, the visible mass in them should not be able to hold onto the stars at their edges since these are whizzing around so fast that the inertial forces should propel them outwards. This is usually solved by adding dark matter to the galaxies arbitrarily to hold them in by force. Instead, MiHsC solves this problem by reducing the inertial mass of the stars. Starting with Newton's inertial second law and his gravity law we get

$$F = m_i a = \frac{GMm}{r^2} \tag{5.5}$$

where m_i is the inertial mass, a is the rotational acceleration (v^2/r) of the star towards the galactic centre, G is the gravitational constant which is just a very small number, M is the gravitational mass within a radius r, and m is the gravitational mass of the star at the galaxy's edge. We can now replace the inertial mass using MiHsC:

$$m \left(1 - \frac{2c^2}{|A|\Theta} \right) a = \frac{GMm}{r^2} \tag{5.6}$$

The acceleration A can now be separated into a mean part that is constant in time that is supposed to represent the steady rotational acceleration of the star around the centre of mass of the galaxy, this can be called a, and a part that varies with time, for example, as the star accelerates into denser galactic spiral arms, close to neighbouring stars, through globular clusters or when it moves outside the plane of the galaxy's disc. This is called a' (a-prime). So

$$m \left(1 - \frac{2c^2}{(|a| + |a'|)\Theta} \right) a = \frac{GMm}{r^2} \tag{5.7}$$

In order to simplify this formula a little, we can divide by m and multiply by $|a| + |a'|$ to give

$$\left(|a| + |a'| - \frac{2c^2}{\Theta} \right) a = \frac{GM(|a| + |a'|)}{r^2} \tag{5.8}$$

You may think that this has helped us not at all, since the formula still looks about as simple as a Rubic's cube, but there is a nice little simplification we can now make and that involves remembering from the last section that MiHsC predicts that there is a minimum acceleration in nature, with no exceptions and that this is equal to $2c^2/\Theta$. This occurs, to recap, because as accelerations reduce the Unruh waves seen by objects become so large that a greater proportion of them do not fit into the Hubble scale and they are disallowed. In principle they cannot be observed and Mach said that if something cannot be observed they should be assumed not to exist. Now at the very edge of a galaxy, the mean acceleration $|a|$ become very small, but still there must be an acceleration above $2c^2/\Theta$ so this acceleration must be provided by the local accelerations $|a'|$. Therefore $|a'| = 2c^2/\Theta$ and in the equation above the second and third terms on the left hand side cancel. So

$$a^2 = \frac{GM(|a| + |a'|)}{r^2} \tag{5.9}$$

This looks a bit simpler. As before, we can say that at the edge of the galaxy the mean rotational acceleration $|a|$ is much smaller than the local acceleration $|a'|$ so we can just ignore the $|a|$ on the right hand side since it is dwarfed by $|a'|$. So

$$a^2 = \frac{GM|a'|}{r^2} \tag{5.10}$$

Now assuming a circular orbit we can say that $a = v^2/r$ where v is the rotation speed and r is the radius so the equation becomes

$$v^4 = GM|a'| \tag{5.11}$$

As we already discussed, the acceleration a' at the edge of a galaxy has to be $2c^2/\Theta$ since MiHsC precludes accelerations lower than that so we need to replace a' with this:

$$v^4 = \frac{2GMc^2}{\Theta} \tag{5.12}$$

Now we have a formula that predicts the outer rotation speeds of galaxies for comparison with those observed and *it has no adjustable parameters*. This is a crucial point since MiHsC then has only one chance to be right, and can't be 'tuned', the values are all known. The value of G has been determined in the Solar System and on the Earth and is well known ($6.6 \times 10^{-11} \, \mathrm{Nm^2kg^{-2}}$). The M is the baryonic mass of the galaxy, so this is the mass we can see and this is determined by looking at how much light is coming out of the galaxy and assuming that the amount of light you get is related to the mass in the same way that the amount of light from the Sun is related to its mass, which is known. This is called the mass to light ratio. The value of the speed of light is now very well known. The size of the observable universe is perhaps the least well known of the parameters, but it can be derived from the Hubble constant with an error of 9% (Freedman, 2001).

What remains is to compare the prediction to the observations, so Figure 5 shows the results. The x-axis shows the mass of the system looked at in solar masses. The mass of the Sun is one solar mass, so a value along this axis of 1.E + 03 means 1000 times the mass of the Sun, so these systems range from 1000 solar masses, which tend to be dwarf galaxies such as are found surrounding the Milky Way, up to spiral and disc galaxies in the middle right up to galaxy clusters up to 1×10^{15} times the mass of the Sun. This is a huge range. The

84   *Physics from the Edge: A New Cosmological Model for Inertia*

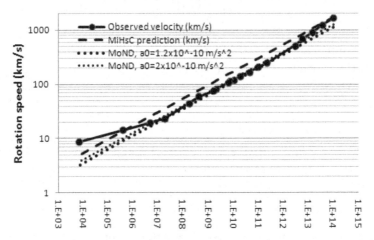

Figure 5. The observed outer rotational velocity for cosmic structures ranging from dwarf galaxies, through gas discs and disc and spiral galaxies and out to galaxy clusters from McGaugh *et al.* (2009) (black circles, solid line) presented in comparison with the predictions of MoND (dotted lines) and MiHsC (dashed line). Both MoND and MiHsC correctly predict the data within the uncertainties in both the data and the models, but MiHsC does this without adjustable parameters.

y-axis, the vertical one, shows the rotation speed of each particular system, and the observation is shown by the solid line. Note that this axis is logarithmic, which is just a complicated way of saying that it has been squeezed smaller in the vertical direction for the large values at the top to get them all on the page. The observations are from McGaugh *et al.* (2009).

The first thing to notice is that the rotation speed goes up as the central mass goes up. This is to be expected: as more matter is present in the centre of the systems, they can possess a greater rotational velocity and still remain stable against the inertial (centrifugal) forces that would like to tear them apart.

The dotted line in the figure shows the prediction of MoND (Milgrom's Modified Newtonian Dynamics) with two different values for its adjustable parameter, $a_0 : 1.2 \times 10^{-10}$ m/s^2 and 2.0×10^{-10} m/s^2. This is done by using the formula: $v^4 = \sqrt{GMa_0}$. MoND works best for the intermediate mass systems, and performs less well on the small dwarf galaxies and the huge galaxy clusters. This might be expected since MoND was tuned, that is the adjustable parameter in it: a_0, was tuned to fit the intermediate spiral galaxies.

The dashed line shows the prediction of MiHsC. This has a 20% error because of the 9% error in the Hubble constant mentioned above and a factor of 2 uncertainty in the stellar light to mass ratio (as suggested by McGaugh *et al.*, 2009). MiHsC overestimates the rotation speed of the intermediate structures, by up to 50%, but this is still in agreement with the data given the large error bars. MiHsC is much closer to the observations for dwarf galaxies and galaxy clusters. MoND is famously unable to model galaxy clusters, but MiHsC can do this.

It is also remarkable that MiHsC performs so well because it does not have any adjustable parameters and so, unlike MoND, it cannot be tuned to fit the data. This is an important advantage for a theory. It is possible for a 'bad' theory to be tuned to fit the data using an adjustable parameter, whereas, if a theory fits without any adjustment, then either it is an example of very good luck, or it is telling us something about nature.

The Flyby Anomalies

So far, I have mostly focused on cosmology and astrophysics, but it is better to test theories on observations closer to home, since they are more reliable and less open to multiple explanations. So I will now talk about the flyby anomalies, show that MiHsC is able to explain some aspects of them, but that the results are mixed.

Fuel is expensive and heavy to launch so interplanetary missions were given a huge boost when Michael Minovitch (1961) proposed the method now called the gravity assist. The film that always helps me to understand this is "Back to the Future". In one early scene Marty McFly is on his way to school on his skateboard and grabs onto a truck which pulls him along for a while. That is what happens with a gravity assist. Say, you want to send a probe into the outer

86   *Physics from the Edge: A New Cosmological Model for Inertia*

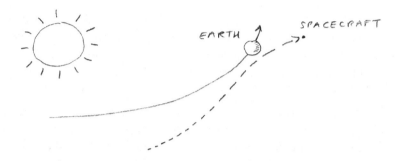

Figure 6.   **An Earth flyby assist.** A spacecraft approaches behind the Earth in its orbit and is pulled along by the Earth's gravity and speeds up, see the lengthening dashes, without the need for expending fuel.

Solar System, to Saturn, but you don't have the fuel necessary to break out of the Sun's gravity well. What you could do is fire the space probe so it arrives at Jupiter at a position behind it in its orbit, and then Jupiter's gravity will pull the space probe along and speed it up, just like the truck pulled Marty McFly along.

The first spacecraft to use this technique was Mariner 10, which used Venus to slow itself down so that it could go into orbit around Mercury.

Very cleverly this is often done using the Earth itself, this has happened now nine times, and the progress of the spacecraft is tracked using laser ranging. To do this they fire repeated laser beams at the spacecraft and wait for the reflection, and since the speed of light is constant in a vacuum, they can work out the distance to the spacecraft as it changes in time and therefore also the speed. They can also work out the speed by sending signals to the spacecraft and looking at the Doppler shift in the reply.

Antreasian and Guinn (1998) first noticed that for these space-craft the speed with which they left the Earth was sometimes slightly different from that expected and this was called the flyby anomaly. The data from these various flybys is shown in Table 1. The first column shows the name of the flyby craft, columns 2 and 3 the incoming and closest approach velocities, columns 4 and 5 the incoming and outgoing latitudes, column 6 the distance of the closest approach (perihelion) and column 7 shows the observed anomalous speed change.

Table 1.   The flyby anomalies. The name of the spacecraft, geocentric velocity at infinity and at perihelion (km/s), the incoming and outgoing declinations (°), the perihelion distance (km) and the observed anomaly (mm/s).

Flyby craft	v_∞ km/s	$v_{\text{perihelion}}$ km/s	Lat_{in} deg	Lat_{out} deg	Perihelion km	dv mm/s
Galileo 1	8.95	13.74	−12.5	−34.2	7331	3.92 ± 0.08
Galileo 2	8.88	14.08	−34.3	−4.9	6681	−4.6 ± 0.08
NEAR	6.85	12.739	−20.8	−72	6911	13.46 ± 0.13
Cassini	16	19.026	−12.9	−5	7553	−2
Rosetta 1	3.86	10.517	−2.8	−34.3	8327	1.8 ± 0.05
Messenger	4.06	10.389	31.4	−31.9	8718	0.02 ± 0.05
Rosetta 2			10.9	18.4	11700	0.27
Rosetta 3			18.5	24.3	15239	0.4
EPOXI 1			5.2	16.8	21943	0
EPOXI 2			17.1	63.9	49835	0
EPOXI 3					30505	0
Juno					7000	?

I remember when I suddenly found out that a new paper had just been published about all this (Anderson *et al.*, 2008). I downloaded it there and then and read it through during my lunch break. The paper was short, but showed that the data they had at the time (the first six columns in the table) fitted a formula like this

$$dv = 3.099 \times 10^{-6} \times v_\infty \times (\cos \varphi_{\text{in}} - \cos \varphi_{\text{out}}) \qquad (5.13)$$

where v_∞ is the hyperbolic excess velocity which is the velocity the spacecraft would have a long way away from the Earth (outside the gravity well), ϕ_{in} is the incident latitude of the spacecraft and ϕ_{out} is the latitude it exited at. They had noticed this pattern in the data: if the spacecraft left the Earth at the same latitude that it approached then there was no anomaly, but if its latitude changed then there was an anomaly.

A good example of a flyby without an anomaly is the sixth shown in Table 1: the Messenger spacecraft which came in at a latitude of 31.4 degrees and left at 31.9 degrees and showed only a tiny speed anomaly.

88   *Physics from the Edge: A New Cosmological Model for Inertia*

Then there is the NEAR flyby which came in at 20 degrees south and left at 72 degrees south and had the largest speed anomaly of 14 mm/s.

I remember reading this paper and finding it breathtaking and I was upset because I couldn't explain it using MiHsC. So, I had a bathroom break and then went over to a lecture room to have a think about it, and there I realised that I could explain it with MiHsC by thinking in a way I had already considered, but hadn't had the data to point out the way.

I took the unusual step of considering all the accelerations seen by the NEAR spacecraft on its flyby. We can simplify things by looking at the spacecraft and its interaction with just one piece of mass within the Earth (see Figure 7). On its equatorial approach the spacecraft sees the acceleration of the mass P away from it towards the Earth's centre as a mutual acceleration (assuming the spacecraft is far enough away that its own acceleration relative to the mass is small). By MiHsC then the wavelength of the Unruh radiation is

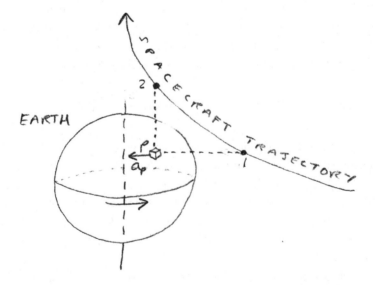

Figure 7.   Schematic showing an Earth flyby where the spacecraft approaches near the equator (1) and leaves near the pole (2). Also shown is an arbitrary mass within the Earth (P).

short and many of the waves are allowed and the inertial mass is close to the gravitational mass.

Now, as the spacecraft leaves close to the pole (see position 2 in the figure) it now doesn't have as much mutual acceleration with respect to the mass within the Earth (P) because the acceleration vector of the mass P is perpendicular to the line joining the craft and the mass, so the mutual acceleration between the craft and the mass is smaller. So the Unruh waves are longer and a greater proportion of them are disallowed by the Hubble-scale Casimir effect, so MiHsC decreases the inertial mass. If we assume that momentum (which is mass times velocity) must be conserved, then the spacecraft, since it has lost mass, must gain in velocity, so it speeds up.

Note also that this data taught me something: the accelerations to consider were not those relative to absolute space, but rather, the mutual accelerations between an object and other objects near to it. Before this I had always thought of accelerations with respect to the background, but this was a move to a kind of thinking closer to Ernest Mach's, as I found out later, and it was not motivated by philosophy, but by necessity: trying to get the model to fit the data.

Now, although I had a hand waving explanation I still needed to show whether the idea predicted the actual numbers correctly, this is the difficult part! Assuming the conservation of momentum for the craft in the reference frame of the Earth we get:

$$m_1 v_1 = m_2 v_2 \tag{5.14}$$

where the terms are: the initial and final momenta of the craft. We can now replace the inertial masses of the unbound spacecraft using the MiHsC formula for inertia:

$$m_g \left(1 - \frac{2c^2}{|a_1|\Theta} \right) v_1 = m_g \left(1 - \frac{2c^2}{|a_2|\Theta} \right) v_2 \tag{5.15}$$

where m_g is the gravitational mass of the craft, or the uncorrected inertial mass. Some algebra implies that

$$v_2 - v_1 = dv = \frac{2c^2}{\Theta} \left(\frac{v_2}{a_2} - \frac{v_1}{a_1} \right) \tag{5.16}$$

This is the new 'jump' in speed (dv) predicted by MiHsC because its mutual acceleration with respect to all the separate masses within the Earth has decreased due to its position closer to the Earth's spin axis.

Now normally, in Newtonian physics, these myriad accelerations are forgotten, and the mass of the Earth is treated as if it was all concentrated at a point in the centre of the planet. With MiHsC instead we need to consider all these myriad accelerations and this is one of the most difficult things to communicate about MiHsC and I am not sure I have ever succeeded in a talk or a paper in getting the idea across clearly. To picture all these interactions one could imagine a line connecting the spacecraft with every Planck mass within the Earth. One then needs to sum up the accelerations (acceleration of the lengths of the lines) for all these interactions and divide by their number to get the acceleration that goes into the equation above.

This often reminds me of a game I used to play with my sister in the car. At night we used to see the streetlights and the way the diffraction of the light through the windows makes it seem as though there is a light beam extending between the streetlight and the car, and we imagined that they were pulling the car along. It is these beams of light I imagine when I think of the flyby setup here.

Also, more recently I go regularly for acoustic scans of the heart since I have a bicuspid heart valve that leaks a bit. I usually have to lie on the bed facing away from the screen, but by twisting my head I can see what is going on there and talk to the technician about it. The heart valve is often visible in black and white, and the computer shows red and blue areas. They show the velocity of the blood towards and away from the sonar. What happens is, the sonar emits a sound which bounces off the material, back to the receiver and from the time interval it takes to come back, and knowing the speed of sound, you can work out the distance to the material it bounced off and plot the sound coming back from a particular layer of the body on the screen.

You can also look at the Doppler shift of the sound. If the pitch of the sound is higher when it bounces back, that means the object (probably blood) is moving towards the sonar and if it is lower, the blood is moving away. In this way the speed of the fluid can be plotted in colour. Always, in looking at my heart one can see a colourful jet of colour which represents the leak back of blood as my heart valve fails to push it all through the valve.

I also think of the flybys in this sort of way. In this case the spacecraft sees accelerations (not speeds) of the Planck masses in the Earth as they spin around. The faster the mutual acceleration the more colour the spacecraft would see, the more Unruh waves are allowed by MiHsC and therefore the inertial mass is higher, and closer to the gravitational mass.

This use of mutual accelerations between two masses, is very satisfying when you look at Mach's point that you should only include in a theory things that you can directly observe. You cannot determine an acceleration relative to absolute space, because you cannot be sure how the space (which is invisible) is accelerating, but you can determine the acceleration between two solid bodies.

Now I won't derive the whole thing here (you can see the Appendix of McCulloch, 2008b if you want to know. It is quite involved because you have to consider the fact that the density of the planet is slightly greater at the centre), but the average acceleration of the masses within the Earth as seen by the spacecraft (the coloured areas) is given by

$$a = \frac{0.07 v_e^2}{R} \qquad (5.17)$$

where the 0.07 comes from the variation of density as you go towards the core, v_e is the equatorial spin velocity of the Earth and R is the Earth's radius. The component of this acceleration seen by a flyby spacecraft at a latitude of ϕ is therefore

$$a = \frac{0.07 v_e^2}{R} \cos \varphi \qquad (5.18)$$

We can now use this expression for the accelerations in the MiHsC formula for dv' and assume that $v_1 = v_2$:

$$dv = \frac{2Rc^2}{0.07 v_e^2 \Theta} \times v \times \left(\frac{\cos \varphi_1 - \cos \varphi_2}{\cos \varphi_1 \cos \varphi_2} \right) \qquad (5.19)$$

We can now substitute values in here. I always like using real numbers. It could almost be thought of as a sign that you are doing something real. In too much of modern physics, the numbers are excluded and people set, for example $c = 1$. I find this to be abhorrent, because it cuts the link between the physics and the real world. It may, as some argue, expose pattern in the theory more clearly, but I think this makes the physics that they do very incestuous. No data can come in from the real world that way. So, I like numbers and they are: $R = 6371$ km, $c = 3 \times 10^8$ m/s, $v_e = 465$ m/s and the Hubble diameter $\Theta = 2.7 \times 10^{26}$ m, so we get

$$dv = 2.8 \times 10^{-7} \times v \times \left(\frac{\cos \varphi_1 - \cos \varphi_2}{\cos \varphi_1 \cos \varphi_2} \right) \tag{5.20}$$

This formula is very similar to the empirical equation derived by Anderson *et al.* (2008) from their flyby data, especially its dependence on the difference in the cosine of the incident and outgoing latitude. Figure 8 shows the observed flyby anomalies in black and those predicted by the MiHsC formula above in grey.

There is the NEAR flyby which has the largest anomaly at 14 mm/s, and MiHsC predicts 10 mm/s. The only anomaly it doesn't predict is the EPOXI-2 and 3 anomalies which were zero and MiHsC predicts jumps of 5 mm/s.

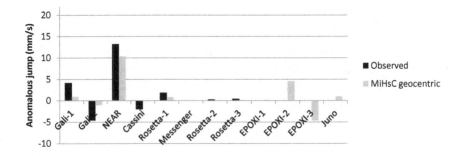

Figure 8.   The flyby anomalies (in mm/s) both observed (black) and predicted by MiHsC (grey). MiHsC agrees with the sign of the anomalies, and with the NEAR anomaly within error, which is the largest one, but it is far from perfect. The EPOXI flybys occurred about ten times further away from the Earth than the others and I would not expect MiHsC to predict them, see the next section on the Tajmar effect.

As I am always reminding my students, when deciding whether or not a theory agrees with the data it is very important to consider what uncertainties there are in the theory and what uncertainties are in the data. If the two agree within these error bars then the theory can be said to be supported by the data. With MiHsC, an error is always incurred by our inaccurate knowledge of the Hubble scale (Θ) giving us an error of 9%. In this particular case there is also an error incurred because of a simplified geometry used in the derivations, for example I have so far neglected the geocentric accelerations between the spacecraft and the centre of the Earth. Maybe if you have the time and inclination you can do a better job.

In 2009 I was invited to attend a flyby workshop at the International Space Science Institute in Bern, Switzerland. There I met one of the discovers of the flyby anomaly, Dr John Campbell, and he gave me the EPOXI data points, which all showed a zero anomaly. That was fine for one of them because MiHsC predicted a very small anomaly, but it was not fine for the EPOXI-2 and EPOXI-3 anomalies, since they were zero and MiHsC predicted a speed jump of about 5 mm/s.

Again, this upset me quite a lot at first, and, as happens with me about twice a week, I began to fear that I was chasing fairies down the garden path, but the interesting thing here was that the perigee of the EPOXI flybys was about ten times further away from the Earth than all the other flybys. What if there is a distance dependence of MiHsC? I realised that there is a distance dependence, not by thinking philosophically about Mach's principle, but rather by looking at another experimental anomaly. This time more down to Earth, though no less controversial: the Tajmar effect. Progress (and hopefully this does represent progress) never lies down a safe path.

The Tajmar Effect

When you are looking at an anomaly in space you can't change the conditions to test your hypothesis, you have no control, so you can never be conclusive. In contrast, a laboratory experiment gives one some control.

94   *Physics from the Edge: A New Cosmological Model for Inertia*

One such experiment was that of Martin Tajmar. Whereas the flyby anomalies were an anomalous motion of spacecraft which were changing their orientation and therefore their acceleration with respect to matter in the spinning Earth, the Tajmar effect is an anomalous motion of a gyroscope close to a ring which changes its rotational acceleration in time (for the details of the experiment see Tajmar *et al.*, 2007, 2008, 2009).

Martin Tajmar and his team used rings made up of various materials (niobium, aluminium, stainless steel and Teflon) which were about 7.5 cm in radius, so about the size and shape of small circular Frisbees and they cooled them in a cryostat down to 5 K (−268°C). They placed accelerometers (devices which measure acceleration) close to the rings (about 5, 10 and 23 cm away) and made sure they were not in any frictional contact with the rings. Then they span the rings both clockwise and anti-clockwise and amazingly they found that the accelerometers measured an acceleration almost as if they were in slight contact with the ring (but this was not so). This acceleration was very small: it was only one in ten billionth the acceleration of the ring itself, but it could be detected. They later used more sensitive laser gyroscopes and obtained the same result.

As they noted this is a little like the Lense–Thirring effect that is predicted by general relativity, where a rotating body pulls around space-time with it, so objects near to a large rotating body are expected to spin around with that body. This effect was apparently measured by Gravity Probe B. However, what Tajmar *et al.* saw is about 20 orders of magnitude (or 100,000,000,000,000,000,000 times) larger than the Lense–Thirring effect expected from such a small ring.

The experimental setup is shown in schematic form by Figure 9. The ring was horizontal and the three laser gyroscopes were placed 5, 10 and 23 cm away directly above the ring's edge.

We can simplify the mathematics of this nicely by forgetting for a moment that we have a spin and just looking in the vicinity of the gyroscope and below it. Here, the ring is simply moving in a straight line below the gyroscope, the curve is small, so, just as we did for the flybys, we can write a conservation of linear momentum equation

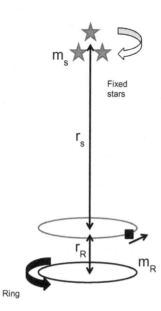

Figure 9.   A schematic of the Tajmar experiment. The figure shows the
rotating ring (bottom), the laser gyroscope (middle) and the fixed stars above.

like this:

$$m_{g1}v_{gr1} = m_{g2}v_{gr2} \tag{5.21}$$

The left hand side is the momentum of the gyroscope with respect
to the ring (hence the subscript "gr", the g is for gyroscope, the r is
for ring) along the direction of the ring's spin before the ring spins,
and the right hand side is the momentum afterwards. This just says
that we are assuming that the momentum doesn't change. If we try
the usual trick of replacing the inertial masses (the "m"s) with the
MiHsC inertial mass we get

$$v_{gr1}\left(1 - \frac{2c^2}{|a_{g1}|\Theta}\right) = v_{gr2}\left(1 - \frac{2c^2}{|a_{g2}|\Theta}\right) \tag{5.22}$$

And rearranging this we get

$$v_{gr2} - v_{gr1} = \frac{2c^2}{\Theta}\left(\frac{v_{gr2}}{|a_{g2}|} - \frac{v_{gr1}}{|a_{g1}|}\right) \tag{5.23}$$

This formula is the same as the one shown above for the flyby anomalies. In that case the initial and final accelerations seen were the mutual accelerations between the spacecraft and the spinning Earth. In this case it is the mutual accelerations between the mass in the gyroscope and all the surrounding matter (including that in the ring) that need to be considered.

First of all, the whole experiment is done in a cryostat cooled down to 5 K, so this means that the temperature-dependent thermal accelerations of nearby atoms should be minimised. We can also say that the accelerations between atoms in the gyroscope and atoms in the solid Earth are small because the experiment is solidly fixed to the Earth's surface.

What accelerations does the gyroscope (which is assumed to be in the same plane as the ground) see? Well, it is fixed to a spinning planet so from its point of view the fixed stars in the sky are spinning around and this is a mutual acceleration, one that Mach would appreciate. The fixed stars are, on average, very far away, but their combined mass is huge.

The apparent spin rate (mutual acceleration) of the fixed stars depends on the latitude you are at. If you are at the equator then the spin is zero, the fixed stars just seem to pass across the sky linearly. If you are at the pole then the stars appear to rotate (spin) around Polaris, the Pole star. If you are in the mid-latitudes there is an intermediate amount of spin. My background is in physics, but also in oceanography, and physical oceanographers know this dependence very well (considering the spin in the same plane as the ground), it is called the Coriolis acceleration and depends on the latitude.

The Coriolis acceleration is best understood by thinking about a missile launched from the equator, say from Equador. The rotation of the equator beneath the missile was zero (around a locally vertical axis), so the missile also has zero rotation. As the missile moves further north or south it is moving into areas that have non-zero rotation. In the northern hemisphere the missile would see the ground beneath it start to spin anticlockwise. A person standing on the ground looking up at the missile would see the missile spinning clockwise, and as it travels along it would appear to turn to the right. In the southern hemisphere everything would be reversed.

At Siebersdorf in Austria, where Tajmar *et al.* conducted their experiment the latitude is 48°N, and, as any physical oceanographer knows, the Coriolis acceleration is fv, where f is the Coriolis parameter (0.0001/s in mid-latitudes) and v is the spin velocity of the Earth at that latitude (which is 311 m/s). This gives an apparent acceleration with respect to the fixed stars of 0.0311 m/s^2. This is also the spin acceleration of the experiment with respect to the stars (since it is on the Earth spinning anticlockwise in the northern hemisphere) but since only the mutual acceleration is important, they are the same. So, in the above formula the initial mutual acceleration seen by the gyroscope is $a_{g1} = 0.0311$ m/s^2.

When the ring was suddenly rotated by Tajmar, there was an acceleration of $a_R = 33$ rad/s$^2 = 2.5$ m/s^2, but in order to apply MiHsC to this case study there is a problem. What is the relative importance for the fixed stars and the ring for the inertial mass of the gyroscope? Are they equally important? Surely not! Surely larger and closer masses should be more important? What I chose to assume was that the importance of another mass for the inertial mass of the first is proportional to the mass over the distance squared, by analogy with gravity. So, to calculate an average mutual acceleration (a_{g2}) for use in the MiHsC formula for inertia we need to write down the acceleration of the heavy but distant stars and the light but close ring, as follows:

$$\bar{a}_{g2} = \frac{\frac{m_s}{r_s^2}a_{gs} + \frac{m_R}{r_R^2}a_{gr}}{\frac{m_s}{r_s^2} + \frac{m_R}{r_R^2}} \tag{5.24}$$

where m_s and r_s are the mass and the average distance away of the fixed stars, and m_R and r_R are the mass and distance away of the ring. We can now put values in for these quantities. The mass of the observable universe is not a very robust figure, but this has been estimated by counting stars to be close to 2.4×10^{52} kg. The average distance away of these stars can be estimated to be the Hubble distance which is 2.7×10^{26} m ($r_s = 2c/H$, where H is the Hubble constant from Freedman, 2001). The mass of the (steel) ring can be estimated from the density of steel ($8000\,\text{kg/m}^3$) and the dimensions of the ring. It had a circumference of $2\pi \times 0.075$ m, a height of 0.015 m and a width of 0.006 m. The distance from the lowest gyroscope to the ring, was 0.0533 m. Using these values we have

$$a_{g2} = \frac{0.33a_{gs} + 118a_{gr}}{118} \tag{5.25}$$

We can therefore forget all about the a_{gs} value since it is small compared with the other term in the formula, and simplify this a little

$$a_{g2} = a_{gr} \tag{5.26}$$

So the final acceleration is simply the acceleration of the ring, and the fixed stars are irrelevant (but they are not irrelevant before the ring is spun because they are responsible for the initially low inertial mass). We can now substitute the values for a_{g1} and a_{g2} into the conservation of momentum formula:

$$v_{gr2} - v_{gr1} = \frac{2c^2}{\Theta}\left(\frac{v_{gr2}}{|a_{gr}|} - \frac{v_{gr1}}{|a_{gs}|}\right) \tag{5.27}$$

The first term in the brackets has the acceleration of the gyroscope with respect to the ring on the bottom, and this is a pretty large quantity when the ring is spinning so fast, so this means that this term is divided by a big number, and so is very small and we can forget it. So we get left with

$$v_{gr2} - v_{gr1} = \frac{2c^2}{\Theta}\left(-\frac{v_{gr1}}{|a_{gs}|}\right) \tag{5.28}$$

Now we can differentiate this change in velocity (dv) with time to find the change in acceleration (da)

$$da = -\frac{2c^2}{\Theta}\left(\frac{a_{gr}}{|a_{gs}|}\right) \tag{5.29}$$

If we swap the a_{gr}, the acceleration of the gyroscope with respect to the ring, with the acceleration of the ring with respect to the gyroscope (a_{rg}) then we have to multiply by -1, so

$$da = \frac{2c^2}{\Theta}\left(\frac{a_{rg}}{|a_{gs}|}\right) \qquad (5.30)$$

This equation predicts that the rotational acceleration of the gyroscope will be in the same direction as that of the ring, just as observed by Tajmar *et al.* in their experiment and will be proportional to the ratio of the acceleration of the ring below it to the acceleration of the fixed stars above it. If we assume first that the acceleration of the gyroscope with respect to the fixed stars is constant then this gives us an acceleration of

$$da = 6.6 \times 10^{-10}\left(\frac{a_{rg}}{0.0311}\right) = 2.1 \times 10^{-8}a_{rg} \qquad (5.31)$$

To compare this prediction with the data, we also need to know how uncertain it is. It is known that there is a 9% uncertainty in the Hubble constant, which leads to a 9% uncertainty in Θ, and 9% of the prediction is about 0.2. So the prediction is

$$da = (2.1 \pm 0.2) \times 10^{-8} \times a_{rg}\, \text{m/s}^2 \qquad (5.32)$$

This is happily in agreement with the acceleration ratio seen by Tajmar *et al.* (2009) which was $(3 \pm 1.2) \times 10^{-8}a_{rg}\,\text{m/s}^2$. All the details can be found in McCulloch (2011a).

That is the meaty part of the description of the way that MiHsC predicts the results of the Tajmar experiment. MiHsC predicts the result very well, and it does this without adjustable parameters. The trick is to think about things in a slightly different way: inertia here is indeed due to matter there, as Mach speculated, but you have to account for things properly.

Nonetheless, there are problems with the experiment. For a start the Tajmar experiment has never been fully reproduced, not even by Tajmar, although maybe this is not surprising since it was done pretty recently, and requires a sophisticated set up, and Tajmar himself has been moving around the world, from Vienna, to South Korea, to Germany. Maybe it will be performed again soon.

In light of the above, an especially interesting experiment would be an attempt to do the experiment in the southern hemisphere, since MiHsC then predicts that the gyroscope would still follow the ring in its rotation, but that in the south it would be the anticlockwise rotation that would be slightly larger. This is because the spin of the Earth (of the stars, actually) should produce a parity violation. So, people in New Zealand, Australia, Southern Africa and South America, etc.: please have a go! Some guys in New Zealand (Graham *et al.*, 2008) did attempt it, and the parity violation in their results does indeed look reversed, but unfortunately their data had error bars much larger than the results, so the result must be discounted (Graham *et al.*, 2008).

When I have been through a bout of mathematics I always like to recap things intuitively, since that is where all theories must eventually be considered, so here goes.

Before its surroundings are cooled the gyroscope sees all kinds of local accelerations due to the heat of the molecules near to it. They are vibrating and accelerating, so the gyroscope sees Unruh waves that are relatively short compared to the Hubble scale, so very few of them are disallowed by the Hubble-scale Casimir effect and the gyro's inertial mass is normal.

Then the surroundings are cooled down to $5\,\mathrm{K}$ in the cryostat and suddenly most of these thermal accelerations cease. The gyroscope is also not accelerating with respect to the Earth, at least at first, so the only acceleration it sees is with respect to the fixed stars. The gyroscope is on the spinning Earth, so it has a mutual acceleration with respect to the fixed stars, but this is small, so the Unruh waves it sees are longer now and a greater proportion of them are disallowed by the Hubble-scale Casimir effect because they don't fit within the Hubble scale. The inertial mass of the gyroscope decreases in this case by about $2 \times 10^{-8}\,\mathrm{kg}$ for every kg of mass because of MiHsC, so presumably it must have rotated slightly with respect to the fixed stars at this point, but this was not monitored by Tajmar.

Suddenly the ring accelerates and the gyroscope sees the Unruh waves produced by that mutual acceleration. These are short, since the acceleration is large, and so fewer of them are disallowed by MiHsC and the inertial mass of the gyroscope goes up. Now how does the system represented by the gyroscope and the ring conserve its momentum (mass times velocity)? The inertial mass of the gyroscope (m) has just gone up, so its velocity has to go down, in the reference frame of the ring. The only way this can happen is if the gyroscope starts to rotate with the ring, so their mutual velocity is reduced. A further complication is that when the gyroscope starts to rotate with respect to the fixed stars, this generates more mutual accelerations that cause a parity violation since a clockwise rotation increases a_{gs} and an anticlockwise rotation decreases a_{gs}.

Something I did not consider in my paper, is that when it starts to move, the gyroscope also starts to accelerate with respect to the matter in the solid Earth, so this should increase accelerations and increase the inertial mass still further. This complication deserves another paper, but I have not yet had time to go into details like this. The art of scientific revolution, if this is what this is, is to focus on things that others will regard as being conclusive proofs. Easier said than done!

There are plenty of embarrassments when one is trying to do something new, and one of these was brought home to me, actually in the town where Einstein has his annus mirabilis: Berne in Switzerland. When I had just published my papers on MiHsC and the flyby anomaly I was invited to attend a week-long workshop on the flyby anomaly.

I was invited back a year later and this workshop was also organised by Professor Claus Lammerzahl of Bremerhaven, who is probably one of the most impressive men I have met in my life: huge, German, professorial, etc., and I think I did disappoint him on this occasion. This was because the night before I was to give my talk on MiHsC to the workshop, I was going through my proof that MiHsC predicts the Tajmar effect (in my first paper on it) and at about 10 pm I noticed a huge flaw in the mathematics in my published paper! I had used completely the wrong reference frame. If you are curious, you can read the first paper I wrote on this — please do not! — and eventually you will realise that, although the intuitive idea is perfectly sound, I am basically talking absolute mathematical rubbish by the end of it.

The next day, I only had half a talk to give: the first part about the flyby anomaly and I then sat down with still another hour to go. Prof. Lammerzahl came over to ask me politely where the rest of my talk was, and I had to tell him that was it. It was extremely embarrassing for me, and maybe I should just have presented the flawed derivation and asked for help, but I didn't. Furthermore, that night I noticed (you can check these things on the arxiv) that lots of people were reading my flawed paper! (Presumably the workshop attendees and their colleagues back home.) For months I had been dreaming about people actually reading my paper, and few people had noticed it. As soon as I realised that the paper was flawed, everyone was suddenly reading it and presumably realising the flaw and writing me off for evermore.

Later that year my paper actually won an award from the journal it was published in, and despite its flaws I do think it deserves it, because although it is flawed mathematically, I still think it is groundbreaking and intuitively correct. It is interesting that when I first submitted it to the journal the maths was right and only got screwed up as I lost track during the review process.

Anyway, I once told Prof. Horne at the University of St Andrews about my error and he made a comment that cheered me up a lot: "Don't worry: these are hard things to think about." I think that is absolutely right, although he did later go on to say that "my maths was a little sloppy", and I think that is right also. In fact, the proof of my sloppiness has been published! Nevertheless, do not let this put you off MiHsC, since I devised it intuitively, which I usually seem to get right. The mathematics always comes later, and although I make frequent mistakes, I keep going obsessively till the mathematics fits my intuition. I wrote another paper to correct the first one, and that also got an award.

The way I introduced the Tajmar effect at the beginning of this chapter was by saying it was "the flyby anomaly writ small" and that it would offer a solution to the erroneous prediction of the EPOXI-2 and EPOXI-3 flybys. If you remember these flybys showed no jump at all, but MiHsC predicted a large one of 5 mm/s or so. I also said that these flybys were much further away than all the others, and if you look at Eq. (5.24) above you will see that the influence of an object on the inertial mass of another one, in MiHsC depends on the square of the distance between them. This is something I have added to MiHsC since thinking about the EPOXI flybys and the Tajmar effect. Now, you may say I am just 'fitting' the theory to the data, but this idea makes perfect sense: effects often decrease with distance, and I have still done it without adjustable parameters. The fit I have suggested (to $1/r^2$) is extremely simple and is analogous to the gravitational effect. What this idea means for the EPOXI flybys is that the EPOXI spacecraft was so far away from the Earth (about 10 times further away than all the others) so its acceleration with respect to the spinning Earth was less important, in its version of Eq. (5.24), and its, relatively constant, acceleration with respect to the fixed stars and the other planets was more important, so there was much less of a jump in speed.

This distance dependence also provides a way to test MiHsC using a Tajmar experiment. It predicts that the effect should decrease with distance away. For example at 1 metre, 5 metres, 20 metres and 56 meters away from the ring the effect of MiHsC should be 0.03%, 0.8%, 11.5% and 50% smaller. The decrease at 20 meters should be detectable, but, admittedly, cryostats this long are hard to find. Maybe the experiment could be done in space? MiHsC also predicts that the anomalous effect should disappear very quickly outside the cryostat, because of the larger thermal accelerations there.

There is another deeper meaning to all this that relates it to Mach's principle. If you remember, Mach considered Newton's bucket experiment and claimed that the water in the bucket was becoming curved, not because it was accelerating relative to absolute space as Newton had claimed, which is a thought thing only, but rather that it was the accelerations relative to surrounding matter that were important, specifically accelerations relative to the fixed stars. Mach wondered what would happen if, instead of rotating the bucket, one could rotate the fixed stars instead and he said that no-one was competent to say how this would turn out (Bradley, 1971). Would the surface of the water curve or not?

Thinking about this using MiHsC makes a lot more sense and enables us to predict what would happen if the fixed stars were rotated. MiHsC predicts that if the bucket itself is rotated the inertial mass of the water will undergo a slight increase as it suddenly rotates relative to the fixed stars, just like the gyro discussed above, and this gain in inertial mass will slow the water down in its rotation. Another way of saying this is that a new (tiny) MiHsC rotation will appear contrary to the rotation of the bucket.

Now, in the case where the bucket is static but the fixed stars are suddenly rotated, then again the inertial mass of the water will slightly increase following MiHsC and the higher mutual acceleration and to conserve momentum, the water will rotate (slightly) with the fixed stars, and this is in the same direction as before: contrary to the motion of the bucket.

This all sounds very philosophical, but I have now discussed several observations that fit the predictions of MiHsC. With the flyby anomalies the results were not very impressive, but as I said in that section, I have not yet been able to model all the accelerations present as the spacecraft flyby. In the flyby case the observed jumps in the speed of the spacecraft can perhaps be explained as variations of the inertial mass of the craft because of the acceleration (spin) of the matter in the nearby spinning Earth. With the Tajmar effect the inertial mass of the gyroscope, as apparent in its motion, is affected by both the nearby spin of the ring and the spin of the distant stars. These experiments show that in cold/low-acceleration environments Mach's principle (inertia here is due to mass there) may have been seen.

The Pioneer Anomaly

Before we leave the tests of MiHsC that I have published, and talk about possible future ones, I would like to talk about the anomaly that, for me, started all this: the Pioneer anomaly. First, I shall talk about the Pioneer mission, based on information given in the comprehensive summary of Anderson *et al.* (2002).

On 2nd March 1972 and the 5th April 1973 Pioneer 10 and 11 were launched from Cape Canaveral. These spacecraft used gravity assists at Jupiter and Saturn to give them the boost in speed necessary to leave the gravity well of the Solar System.

The crucial characteristic of the Pioneer spacecraft is that they were spin stabilised (spinning), rather like bullets, and this was done as it is for bullets so that they would be more likely to travel in a straight line, without the need for messy rocket course corrections. At launch they were travelling at 4.28 and 7.8 revolutions per minute respectively. This means that for the Pioneer spacecraft very few orbital corrections, and engine firings, were needed and so the accelerations of the craft could be determined to an accuracy of 10^{-10} m/s^2. In contrast the Voyager spacecraft were not spin stabilised and required frequent rocket propelled course corrections the results of which are difficult to predict, so there is a lot of noise in the trajectory data and accelerations cannot be determined so accurately. So, to quote Anderson *et al.* (2002) directly: "the Pioneer spacecraft represent an ideal system to perform precision celestial mechanics experiments". This was partly deliberate, since it was felt that small perturbations in the spacecraft trajectory would reveal the presence of small bodies in the outer Solar System. The best experiments are those where you can reduce the number of effects acting on a system down to one, so you can learn everything about that one effect. That is why the Pioneer craft make such a beautiful experiment: because they are in space the effects acting on them are minimised. Away from the Sun there is just gravity...

...well, not quite! One of the main problems early on in this celestial mechanics experiment was solar radiation pressure. Photons streaming out from the Sun carry momentum, and bang into the spacecraft pushing it and this did have a detectable effect on the trajectory within 20 AU, but it gradually reduced so that by the time the craft had passed the orbit of Saturn it could be neglected. Then, oddly enough, Anderson *et al.* (1998, 2002) found a systematic extra acceleration pushing the Pioneer craft back towards the Sun that was

$(8.09\pm0.2)\times10^{-10}$ m/s^2 for Pioneer 10 and $(8.56\pm0.15)\times10^{-10}$ m/s^2 for Pioneer 11.

The analysis published by Anderson *et al.* (2002) showed that the anomaly could not be due to solar radiation pushing on the craft, the solar wind pushing it (both in the wrong direction), the effect of the solar corona on the radio waves used to communicate with the craft, electromagnetic Lorentz forces acting on a charged spacecraft, the gravity of the Kuiper belt, the drift of clocks on board the space-craft, or the effect of the antennae used by the Deep Space Network "drooping" due to gravity (!). Also considered was the radio beam reaction force. When the Pioneer spacecraft beam a signal, a radio wave, at the Earth, there is a small recoil and it was shown that this is also too small and in the wrong direction to explain the anomaly (it was about 1.1×10^{-10} m/s^2).

Probably the most debated explanation of the Pioneer anomaly is the work lead by Turyshev *et al.* (2012) that the Pioneer anomaly is due to heat radiation from the Radio-Thermal Generators (RTGs) bouncing off the spacecraft antennae and pushing it sunwards. They and the folk at the US Planetary Society have done an excellent job raising funds to recover all the Pioneer data and digitising it into a modern format. This work is essential, since good observations are crucial. I also appreciate all the work they have put into analysing this data, but I have some criticisms of their thermal explanation:

(1) As they say in their 2011 paper (on page 4) the Pioneer data is too noisy to prove whether there is a decay with time in the anomaly or not, and a thermal explanation can't be supported without a proven decay since the RTGs output declines signifi-cantly with time.

(2) The half-life of the decay that best fits their thermal model is 28.8 or 36.9 years whereas the half-life of the plutonium on board is 87 years.

(3) Their predicted anomaly is at its largest in the inner Solar Sys-tem where there was no Pioneer anomaly. They have got around this by proposing that there was an exactly cancelling push because the sunward side of the craft was warmer due to sunlight, but looking at their Fig. 2 from their latter paper this does not

exactly cancel the onboard thermal effect, so I guess they had to adjust the momentum flux from photons close to the Sun that was originally assumed by Anderson *et al.* (1998).

(4) Anderson *et al.* (2002) said that since most of the heat from the RTGs was radiation from fins whose flat surfaces were not pointing at the antenna, only their narrow edges, only 4 W of power, could have hit the antenna, leading to a maximum acceleration of only 0.55×10^{-10} m/s^2.

(5) Since the power radiation Q is proportional to the fourth power of the temperature, I would like to see the temperature errors they get, since any errors from this source would be hugely magnified.

(6) More generally, I always find it difficult to accept a paper when a very complex and unrevealed process (over 3000 finite elements were used in their thermal model, not detailed in the paper) and with two fitting parameters, is used to get to a previously known answer, and no experiment is suggested that might unambiguously test it against rivals. I dislike dark matter for similar and more extreme reasons.

(7) The details of their model have not been published so their model is not falsifiable. To make it falsifiable I would like to see them present a simplified order of magnitude calculation so others can reproduce what they have done with the thermal model, on paper.

Having said all this, I cannot prove that they are wrong, as I said their research is unfalsifiable since they have not published the details. I simply suspect that they are, since they seem to be working with the same kind of attitude that most of modern theoretical physics is: that physics is complete, and so if the data doesn't agree with the theory you can manipulate the data by inventing invisible matter, or use increasingly tortuous methods to make the data fit the theory. I am sure Turyshev *et al.* have their hearts in the right places, but this is an attitude that is endemic to all modern theoretical physics and it needs to be fixed, possibly at the physics degree level, since passing exams encourages a parrot mentality. This is not as bad in physics as in say, history, where facts are learned parrot fashion, but what is learned in physics is the model, and exams test whether students use the model properly. What exams do not do is test the student by giving them raw data and asking them to invent a new model.

Anyway, back to the Pioneer anomaly. It is very easy to explain it with MiHsC, since, as shown above the equation of motion for MiHsC is as follows:

$$a = \frac{GM}{r^2} + \frac{2c^2}{\Theta} = \frac{GM}{r^2} + (6.7 \pm 0.7) \times 10^{-10}\,\text{m/s}^2 \qquad (5.33)$$

The observed accelerations was $(8.74 \pm 1.33) \times 10^{-10}$ m/s^2 (an average for both of the craft, from Anderson *et al.*, 2002) and so MiHsC agrees with it. Note that MiHsC has no adjustable parameters, is a very simple model and also makes sense physically.

To explain intuitively: when the Pioneer craft were launched they already had the anomalous accelerations predicted by MiHsC, since it is simply added onto the equation of motion to give an extra acceleration for every object no matter what its state of motion is. Initially, because all the other accelerations were so large, this extra acceleration was hidden in the noise, but as the Pioneer trajectory moved out into the Solar System the other forces acting became smaller and the noise vanished, so that the MiHsC acceleration stood out.

The idea here is that MiHsC acts to decrease the inertial mass of the low acceleration spacecraft slightly below the gravitational mass, so the gravitational force acting on the craft from the Sun is the same, but since the craft's inertial masses are slightly less, they can respond slightly more to the Sun's gravity by accelerating more than expected towards the Sun, as they did.

However, if the inertial mass of everything at the edge of the Solar System is lower than the gravitational mass such that there is always an anomalous acceleration towards the Sun, then what about the planets? They orbit at the same distance from the Sun where the Pioneer craft showed up an anomaly. The anomaly exists further in, but is obscured by other things. Why don't the planets show anomalies? The answer to this could be that the slightly lower inertial mass of the planets can be hidden by errors in our estimates of their gravitational masses.

As this chapter has shown, MiHsC has been developed by looking at controversial and messy anomalies from the real world. Despite this, or rather because of it, MiHsC does successfully predict the very well observed anomalies of cosmic acceleration and galaxy rotation. The remaining problem is that other, less elegant, theories can also explain these, so a more unambiguous test is needed. In the next chapter some will be proposed.

Chapter 6

Future Experimental Tests of MiHsC

One should always be able to test theories. I would go further and say that if you invent a theory that cannot be tested, you should forget the theory. This was the basic principle of science outlined by Francis Bacon in the 17th century. It inspired the Royal Society, which had as its motto "Nullius in verba" which means "Words mean nothing", or "Take no-one's word for it" and Newton who was a member of the Royal Society was very good at testing his theories.

This attitude led to three centuries of quick scientific advances from Newton to the 1970s, but unfortunately, in recent times, perhaps because of the stellar influence of the later Einstein, who put his faith more and more into mathematics despite a lack of success, the Baconian attitude has died out in mainstream 'theoretical' physics. (It is still alive at places like CERN). Theorists seem to be copying the failed later Einstein, not the earlier successful one. This has given rise to all sorts of untestable theories such as string theory, supersymmetry, loop quantum gravity, dark matter and millions of pounds of funding has been pumped into the black hole of these untestable theories.

I do not wish to make the same mistake, so this section is devoted to discussing experiments that might be able to test MiHsC. Over the past few years I have thought of many different experiments, but since I am not an experimentalist myself and have very little experience of it, I have been unable to properly judge them, and it may be that you are in a better position to.

A Spinning Disc with a Weight Above

In science as elsewhere, to make progress it is important to allow your passion, or curiosity, to direct you, and to rely on your own thinking, and not be put off by the group feeling of the crowd, since the collective intelligence of crowds is usually close to zero.

Having said this, there is a danger in freethinking like this, since humans can make errors, and also if you do freethink the crowd doesn't like you very much, but I think it is worth taking a risk now and then, since new discoveries are never made by committees. Nowhere were the dangers of freethinking in today's climate more exposed than in the case of Eugene Podkletnov's "antigravity" experiment.

In 1991 Eugene Podkletnov and his team, were working at Tampere University of Technology in Finland. They were levitating supercooled superconducting rings about 12 inches in diameter using strong AC magnetic fields and spinning them at 5000 rpm. One of Podkletnov's co-workers happened to be smoking, when he shouldn't have been — so much for the rules —, and they noticed that the smoke rose up in a column above the spinning discs (Cook, 2001). When they investigated this by suspending magnetic and non-magnetic masses above the discs they noticed that they lost a fraction of a percent of their weight, and if the spin of the discs increased so did the effect, up to a maximum weight loss of 2%. The effect was not due to a momentum transfer from moving air since these changes persisted when the masses were encased in glass, and the phenomenon was independent of the masses' composition. It was not a magnetic effect since it persisted when a metal screen was placed between the mass and the disc. A schematic is shown overleaf.

After Podkletnov published these results in a reputable journal, Physica C, and after he had had another paper on it accepted by the Journal of Physics D: Applied Physics, a newspaper report written by Robert Matthews and Ian Sample appeared in the Guardian saying that Podkletnov had discovered a kind of "antigravity". For some reason there was a huge fuss about this word, and suddenly Podkletnov found himself sacked by his university. It seems that finding an effect that cannot be explained by mainstream physics is a capital crime in physics circles. It should not be. It should be appreciated that the experiment might be an error, but also it might be a genuine anomaly: a way to learn more than we know now about the world. People should be encouraged to take experimental risks like this rather than being beaten around the head if they so much as challenge accepted wisdom. Without these risks, progress is impossible.

I have been fascinated by the Podkletnov results ever since about 2002, when a friend sent me an online article about it, and I finally published a paper in 2011, testing MiHsC on the Podkletnov results. I had been warned by some physicist friends not to do this, since they said I would be labelled as a crank, but I published it anyway, since I am not paid by any physics department so I can do what I want. This has the disadvantage that I am on the bread line, but it has the advantage that I don't have to listen to the crowd.

There have been some consequences of me publishing something mentioning Podkletnov in it. For example, my university received one letter of complaint from the public saying that I should not be allowed to research experiments like this, and my department head, Professor Neil James, after checking me out, wrote a brilliant letter back to this chap saying that a university researcher should have the freedom to investigate whatever they want. I could have kissed him, but didn't! It also had the consequence that my papers are now put into "general physics" on the arXiv, a sort of naughty spot for errant researchers, but it doesn't really bother me (although at the moment my latest paper has been held by the arXiv for several months since

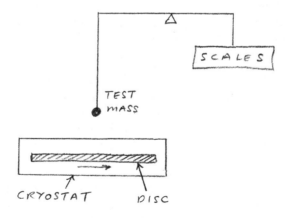

Figure 1. **Schematic showing the setup of the Podkletnov experiment.**
A superconducting disc was placed in a cryostat, cooled to 77 K, levitated and spun. A test mass suspended above the cryostat lost 0.06–2% of its weight.

they can't classify it, whereas it was published in the journal within three weeks). Luckily, journal reviewers are still accepting my papers. My leap off the deep end also had a more important, to me, positive effect that for example, the Lifeboat Foundation invited me to join them, and I did. They are dedicated to saving mankind by helping us (among other things) colonise other worlds, and this latter is a cause I am proud to help them with. There is a lesson here: take risks and you may embarrass yourself in the eyes of the ignorant, make mistakes and probably lose money too, but on the upside, at least you will be noticed by interesting people.

Anyway, the results I got from comparing Podkletnov's results with MiHsC were mixed. Intuitively, and the reason I looked at it, was that I could see how MiHsC might apply to this experiment. There is a mass suspended near a cold environment, so the accelerations it sees are due only to its motion with respect to the fixed stars, and this is small, so initially Unruh waves are disallowed and the inertial mass reduces below the gravitational mass. Then suddenly you switch on an AC magnetic field which has the effect of vibrating the superconductor. The suspended mass now sees larger mutual accelerations and gains inertial mass. Calculating this acceleration in the same way as was done for the Tajmar effect (Eq. (5.30)), shows that the vibration of the disc should, via MiHsC, cause an increase in the inertial mass of the test mass and therefore it should become less sensitive to the acceleration of gravity. Specifically, given the acceleration of the disc by the AC magnetic field, MiHsC predicts the test mass should accelerate upwards (against gravity) by 0.03% of g. The weight loss observed by Podkletnov was 0.06% of g.

However, when Podkletnov spun his discs the effect increased up to between 0.3% and 0.6% of the weight, and I have not been able to explain this using MiHsC (it may be possible, but I don't know enough about the accelerations caused by a superconducting disc that is both vibrating and spinning).

The Tajmar experiments, done over ten years later, were similar to the Podkletnov experiment, but used spinning normal metal rings rather than levitated superconducting discs. The rotating ring of Tajmar is preferable in some ways since a simple rotation instead of a complex vibration provides a 'known' acceleration that can be plugged into

MiHsC and the predicted mass loss can be tested. The disadvantage of the Tajmar experiment was that the accelerations were very small so the effect needed a sensitive laser gyroscope to detect it.

The experiment I would like to suggest combines the best of both experiments. A Tajmar-type rotatable ring is placed in a cryostat, and a Podkletnov-type mass is suspended above and outside the cryostat, isolated from any frictional connection to the ring. When the cryostat is cooled, the change in the weight of the mass due to the cooling is noted. This may not be due to MiHsC. Most of this change will be due to changes in buoyancy, air flow due to the huge change in temperature, but once this is done, any subsequent effects should be due to the ring's spin.

To see the effect of MiHsC, we need then to spin the ring at a few thousand rpm. To calculate the change in acceleration (da) this will cause for the suspended mass we can use the same formula that we used for the Tajmar experiment:

$$da = \frac{2c^2}{\Theta} \frac{a_{mr}}{a_{ms}} \tag{6.1}$$

where a_{mr} is the acceleration of the suspended mass with respect to the ring after it spins (the spin-acceleration of the ring) and a_{ms} is the original acceleration of the mass with respect to the fixed stars because the Earth is spinning (the size of this will depend on the latitude the experiment was done at). We can work out the acceleration of the ring as $a = v^2/r$, where v is the rotation speed of the edge of the ring and r is its radius, or $a = 4\pi^2 R^2 r/3600$, where R is the rotation rate in rpm. The results of applying MiHsC to a ring of radius 5 cm (Eq. (6.1)) are shown in Table 1. (The formula is $da = 7.24 \times 10^{-12} \times R^2 r/a_{ms}$.)

Since this acceleration is an attempt by the ring + mass system to conserve momentum, the change due to MiHsC will be measured as an extra acceleration upwards, and so a loss of weight. This anomalous effect should increase as the acceleration of the ring increases, and towards the pole as the acceleration relative to the fixed stars (a_{ms}) decreases.

The experiment should also be conclusive since if MiHsC predicts correctly the change in weight due to different accelerations of the

116   *Physics from the Edge: A New Cosmological Model for Inertia*

Table 1.   The loss of weight predicted by MiHsC for a test mass suspended over a supercooled ring of radius 5 cm with a spin rate as shown in column 1, at the latitude of Plymouth, UK.

Spin rate Rpm	Acceleration of ring m/s^2	Acceleration of bob m/s^2	Acceleration of bob % of g
3000	4935	0.00015	0.0016
10,000	54,800	0.0015	0.018
100,000	547,556	0.176	1.76
753,994	311,541,808	9.8	100

ring, different directions of spin, different weights of the ring, it will be hard to argue with the conclusions. MiHsC predicts that the effect seen in the experiment should be maximised with a wider ring, a faster spin and a higher latitude (or doing the experiment in a frame that follows the Earth's rotation).

Another possible effect in this experiment will be the change in the centrifugal force on the suspended mass. As the inertial mass increases this should increase by a factor of 10^{-8}, but the gravitational force downwards should not increase. Therefore the mass should tend to move upwards. Since the centrifugal force upwards at the equator is about 1/365th of the gravitational force, the net change in the weight of the suspended mass should be $5.3 \times 10^{-8} \times 1/365 = 1.4 \times 10^{-10}$ kg. This effect is smaller, but it would be interesting to consider what might happen if the inertial mass could be boosted in this way by 365 times — an object at the equator would lift off!

One oddity noticed by Podkletnov and his team was that the anomaly they saw extended upwards apparently in a column, since they also measured weight losses for suspended masses in the floor above their cryostat. There is a way to explain this using the extension of MiHsC using local horizons that I published in 2013 (McCulloch, 2013). Since the acceleration of an object is related to the distance to its Rindler horizon $a = c^2/d_R$ one can write Eq. (6.1) in terms of Rindler horizons, like this

$$da = \frac{2c^2}{\Theta} \frac{d_{\text{old}}}{d_{\text{new}}} \qquad (6.2)$$

Instead of looking at changes in local accelerations, this equation shows that the anomalous accelerations can be thought of as being due to changes in the closeness of Rindler horizons. If the distance to the Rindler horizon halves, then an anomalous acceleration appears that is twice as large as the minimum acceleration of MiHsC. In the case of the Podkletnov experiment, the sudden acceleration of the disc pulls the Rindler horizon in from initially 3×10^{18} m to 9×10^{11} m and the closer horizon would exert a greater pull upwards on any objects close to the disc.

This emphasis on Rindler horizons also points out an alternative method of effecting dynamics. Rindler horizons are boundaries of knowledge, so information from behind them should not be available to the accelerated objects. How about gravitational information then? Could you hide a gravitational source behind a Rindler horizon?

For example, Eq. (6.2) can be rearranged to predict the rotation rate (R, in rpm) of a ring (radius 0.1 m) that is required to bring the Rindler horizon seen by material in it closer than the Sun. Using known values for the speed of light and the distance to the Sun (1.5×10^{11} m) this rotation rate is

$$R = \sqrt{\frac{900c^2}{\pi^2 r d_s}} = 23{,}422 \, \text{rpm} \tag{6.3}$$

The gravitational attraction of the Sun is $0.006 \, \text{m/s}^2$ or 0.06% of g. Since the Sun is about half a degree wide in the sky only an area of 1/720th of the ring would feel the disappearance of the Sun so the actual acceleration would be lower than this.

How about the Earth's gravity? Could we block this out? The distance to the centre of the Earth is 6300 km so the acceleration needed would be

$$a = \frac{c^2}{6300000} = 1.43 \times 10^{10} \, \text{m/s}^2 \tag{6.4}$$

This is a huge acceleration and probably not achievable by spinning a ring. However, Nano-Electromechanical Systems (NEMS) have nanopendulums that do produce accelerations of this size. If one were to weigh one such pendulum while it was still and then again when

118   *Physics from the Edge: A New Cosmological Model for Inertia*

the pendulum was accelerating then the pendulum should not be able to see the Earth's gravity, for at least part of its swing, and this could be detectable.

Huge accelerations could also be made by accelerating electrons in superconductors and then decelerating them in the vertical (downwards) direction using ordinary conductors physically attached below. This is another way of thinking about the Podkletnov experiment which had just this setup and may explain the 'gravity shielding' that they saw.

A third way to get these huge accelerations would be to use electrons accelerating around gold nanotips, as done by Beversluis *et al.* (2003) and commented on by Smolyaninov (2008). The acceleration of the electrons over the curved surface of the nanotip produces an acceleration of $c^2/r = 9 \times 10^{22}$ m/s^2. If a whole array of these nanotips was set up then maybe they would generate a Rindler horizon that would affect nearby test masses.

A final thought before we leave the Podkletnov effect. When the disc is spun the inertial mass of the test mass increases by only a very tiny amount: about 10^{-8} kg per kg. How then does this produce an acceleration upwards that is 0.06% of g? We can see this using a thought experiment shown in Figure 2. We have a disc in freefall

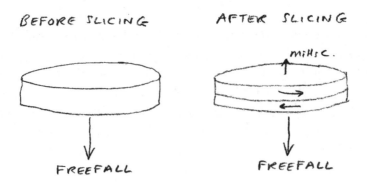

Figure 2.   **Thought experiment.** A disc is in freefall (left). Now slice it horizontally and spin the two halves relative to each other preserving the overall angular momentum. The inertial mass of the two halves increases and to do this you need to overcome a force. Force must be conserved, and a force appears external to the system of just the same size.

on the left. Now slice it horizontally with a laser into two parts and spin them in opposite directions so they are mutually accelerating. MiHsC says that the inertial mass goes up by a fraction dm:

$$\frac{dm}{m} = \frac{2c^2}{a\Theta} \tag{6.5}$$

where a is the spin acceleration. To keep this mutual acceleration going despite the extra inertial mass, a force has to be added which is equal to $F = dm \times a$. Now this force (F) cannot be produced out of thin air. Force should be conserved, so a force must appear external to the system to balance this internal force

$$F_{\text{ext}} = F_{\text{int}} = dm \times a = ma' \tag{6.6}$$

where a' is the anomalous external acceleration produced by the external force. Therefore

$$a' = \frac{dm}{m} a \tag{6.7}$$

In the Podkletnov experiment a is the mutual acceleration of the disc and test mass which (due to the AC magnetic field and the spin) is close to $10^5 \, \text{m/s}^2$. So the anomalous acceleration for a 1 kg mass is

$$a' \sim \frac{10^{-8}}{1} \times 10^5 = 10^{-3} \, \text{m/s}^2 \tag{6.8}$$

This is an attempt at a deeper understanding of what is going on in this case.

The CERN Accelerator and Radio Waves

One of the unique characteristics of waves, and Unruh waves too, is that they can either constructively or destructively interfere with each other. Figure 3 illustrates this. If there are two waves of equal amplitude or height, one with the characteristics of wave A and one like wave B, and they meet up, then where there is a peak in wave A there is a trough in wave B, they are said to be out of phase, and they cancel out. The final result is no wave at all. On the other hand, if wave A meets wave B such that where there is a peak in wave A there is also a peak in wave B, then they will constructively interfere with

120   *Physics from the Edge: A New Cosmological Model for Inertia*

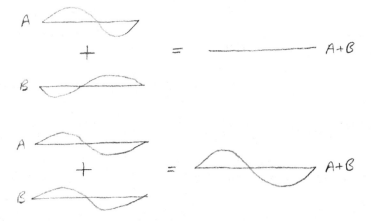

Figure 3.   Schematic showing the destructive interference of two waves out of phase A and B (above), and the constructive interference of two waves in phase (below).

each other and the resulting wave will be twice as big. Is it possible to interfere with Unruh waves? It is difficult if they are as big as the galaxy since we do not know yet how to create waves that long, electromagnetic or otherwise. The wavelength of the Unruh radiation depends on the speed of light (c) and acceleration (a) as

$$\lambda = \frac{4\beta\pi^2 c^2}{a} = \frac{7.11 \times 10^{17}}{a} \tag{6.9}$$

So an object accelerated at $0.0311\,\mathrm{m/s^2}$ would experience Unruh waves of wavelength $2.3 \times 10^{19}\,\mathrm{m}$. An object accelerated at $9.8\,\mathrm{m/s^2}$ would see wavelengths of about $7.3 \times 10^{16}\,\mathrm{m}$. The trick is to produce an acceleration so large that the Unruh wavelengths seen are short enough to be manipulated by present technology. There are some very highly accelerated systems that might fit the bill.

One that I considered as an ideal and cheap experiment for a while was sonoluminescence (described accessibly in Crum, 1994) wherein a collapsing bubble within a fluid undergoes huge accelerations, and oddly emits light by a mechanisms that is not well understood. It may be that this light is Unruh radiation produced by the high acceleration, at wavelengths short enough for us to see

(Schwinger, 1992), but this system does not make a very good experiment because water absorbs most of the radiation emitted making it difficult to draw firm conclusions and the actual maximum acceleration is not well known.

As discussed in the previous section NEMS and nanotips also can produce huge accelerations.

Another man made system that produces huge accelerations is the particle accelerator at CERN (European Organization for Nuclear Research) near Geneva on the French–Swiss border. Here particles are routinely accelerated around the 1 km diameter ring at speeds close to the speed of light. This means that the accelerations produced are

$$a = \frac{v^2}{r} = \frac{(3 \times 10^8)^2}{500} = 1.8 \times 10^{14}\,\mathrm{m/s^2} \qquad (6.10)$$

Putting this acceleration into Eq. (6.9), the Unruh wavelengths seen by these particles should be

$$\lambda = \frac{7.11 \times 10^{17}}{1.8 \times 10^{14}} = 3950\,\mathrm{m} = 3.95\,\mathrm{km} \qquad (6.11)$$

Electromagnetic waves of this wavelength can be produced by our present technology. These are long radio waves. So a possible experiment would be to fire radio waves of this wavelength at the particles being accelerated at CERN to see if we can interfere or modify their inertial mass.

Also since the particles are travelling close to the speed of light, special relativity would have to be taken into account. For example, if the particle was travelling at $0.9c$ (0.9 times the speed of light) then the Unruh waves seen by it would be lengthened in the direction of travel from 3.95 km to about 8 km, so radio waves of this wavelength would instead have to be applied. If MiHsC is right, then the result should be a variation in the inertial mass of the particle, which may be detectable from changes in its trajectory or speed. These changes would be rather sensitive to the phase difference between the Unruh waves and the applied electromagnetic ones.

NEMS Accelerations and Radio Waves

A couple of years ago a Dr. Davide Iannuzzi, from Holland, read my paper proposing the CERN experiment that I have just discussed, and wrote to me suggesting an experiment using the Nano-Electrical Mechanical Systems that he was using. He later decided not to pursue this research because MiHsC's disagreement with the equivalence principle was 'a little scary' but his suggestion was a good one. These NEMS are tiny nano-scale pendulums, and they apparently experience accelerations as high as $4 \times 10^{11}\,\mathrm{m/s^2}$. So the Unruh radiation they see should have a relatively short wavelength as follows:

$$\lambda = \frac{4\beta\pi^2 c^2}{a} = 1.78 \times 10^6\,\mathrm{m/s^2} \tag{6.12}$$

Radio waves of this length, Extremely Low Frequency (ELF), can be generated, and are generated routinely by PIGs (Pipeline Inspection Gauges). So it is certainly possible that radio waves of this wavelength could be fired at the NEMS. Given their wave nature the applied waves may enhance or cancel the existing Unruh waves seen by the pendulum (constructive or destructive interference) therefore boosting or reducing its inertia. Therefore, the motion of the pendulum may be affected in an irregular manner. With more care it should be possible to tune the applied radiation to produce more regular results. Also, the effect will be asymmetric since the applied radio waves will have a different apparent wavelength depending on the relative direction of the pendulum's swing.

To estimate the strength of radio waves we need to apply, we need to start from the mass of a typical garden variety NEMS, which is $2 \times 10^{-16}\,\mathrm{kg}$ (Meyer *et al.*, 2005). Since $E = mc^2$ this inertial mass is equivalent to 18 joules of energy. This energy appears in the pendulum's accelerated frame of reference, as Unruh waves, but cannot be seen by an outside observer. So if we can direct 18 joules of radio wave energy at the pendulum we can potentially dampen or enhance its inertial mass. The focusing of radio waves could be achieved using a parabolic dish, or perhaps a metamaterial. One such metamaterial is discussed in Ehrenberg *et al.* (2012). This one is able to focus microwave radio waves, so would be no use for waves kilometers long, but it is able to focus the waves onto a small volume the size of a molecule and this field is rapidly developing.

Deviations from the Stefan–Boltzmann Law

Part of MiHsC is the assumption that Unruh radiation of very low temperature is weaker than expected because a greater proportion of the longer Unruh waves are disallowed by the Hubble-scale Casimir effect. It is logical to extend this to the temperature of any object, so the radiation emitted by a very cold object should be slightly less than expected from the Stefan–Boltzmann law.

The coldest temperature achieved so far is 100 pK (pico kelvin) at the Helsinki University of Technology (Knuutila, 2000). Using Wien's law, an object this cold would have a peak radiating wavelength of $\lambda_m = 3 \times 10^7$ m, so by analogy to Eq. (5.2), the energy of the black body radiation spectrum (E) would be modified to E', as follows

$$E' = E\left(1 - \frac{\lambda_m}{4\Theta}\right) = E(1 - 2.7 \times 10^{-20})\,\text{J} \qquad (6.13)$$

I don't know if variations in radiated energy as small as this can be detected.

Discretised Variations in Inertial Mass

One of the assumptions of MiHsC is the Hubble-scale Casimir effect, and I have modelled this rather simply by assuming that the proportion of allowed wavelengths decreases linearly as the peak wavelength of the Unruh spectrum increases towards the Hubble scale. This is correct, on average, but if you look closer, it is too crude a model. For certain accelerations, the waves at the peak of the Unruh radiation spectrum (the most common waves) will fit exactly within the Hubble scale, so the inertial mass will be greater at this acceleration. At other accelerations, the waves at the peak, will not fit exactly within the Hubble scale and the inertial mass will be lower for this acceleration.

This implies that, for example, as you consider greater and greater radii in a galaxy you will pass through some radii with stars at the resonant accelerations and these will have greater inertial mass, so will be less attracted by the galactic centre and will move outwards, and some stars in between will have lesser inertial masses, so they will be more greatly attracted by the galactic centre and will move in. The consequences of this are not easy to predict without running a full galaxy model (something I would like to do), but one possibility is that the stars in galaxies should show some tendency to congregate into concentric circles. This tendency will be most apparent at the edges of galaxies where the number of allowed wavelengths is fewer, so the difference between a resonant case, and a non-resonant case will be greater.

Drop Tower Experiment

At the University of Bremen in Germany they have a drop tower that has a fall height of 110 metres. When you drop an object down it, then according to MiHsC it should fall with an acceleration of

$$a = \frac{GM}{r^2} + \frac{2c^2}{\Theta} \tag{6.14}$$

This implies that when you drop something it should arrive at the bottom a little earlier than expected if MiHsC is right. The difference in distance travelled in equal time is $s = \frac{1}{2}at^2$, which for an extra acceleration of $6.7 \times 10^{-10}\,\mathrm{m/s^{-2}}$ and a fall time of $4.74\,\mathrm{s}$ is 7.5 nanometres. If you dropped two objects down the tower, as in Galileo's famous experiment, then MiHsC predicts that they should both fall at the same rate since term 2 on the right hand side of Eq. (6.14) is independent of mass, but both should fall slightly faster than expected.

GPS Satellite Trajectories

GPS (Global Positioning System) satellites are ideal test objects for new dynamics like MiHsC since they orbit at 20,000 km where the atmosphere is thin so their trajectories are less affected by friction which is difficult to model and so fundamental anomalies are easier to see.

In MiHsC the inertial mass of a body is affected by the mutual acceleration between it and nearby matter. This means that as GPS satellites travel nearer to the pole, and the Earth's spin axis, the mutual accelerations between them and the mass of the Earth are lower so their inertia should decrease and they should speed up by a few mm/s (a similar speed-up has already been seen in flyby spacecraft, as discussed in Chapter 5).

Also, in freefall, MiHsC reduces the inertia of objects very slightly, so the satellites should fall in their orbit slightly more than expected (given air drag) giving an extra Earthward acceleration of 6.7×10^{-10} m/s^2. This implies that the GPS satellite should drop by 1 metre for every 12 hours of their orbit. An anomaly of just this size (10×10^{-10} m/s^2) has recently been seen by Ziebart *et al.* (2007) but it is difficult to prove that it is not a mundane (frictional) effect.

Summary of Experiments Suggested

Table 2 lists the various possible experiments that I have suggested to test MiHsC with an attempt to quantify with numbers (1, 2, and 3)

Table 2. This table is a list of the various possible experiments that have been suggested and an attempted quantitative measure of how conclusive (column 2), how practically easy (column 3) and how cheap (column 4) it would be for each of the tests. Column 5 shows the overall score for all of the proposed tests and shows that the disc with a weight above it would be most conclusive, easiest and cheapest, as far as I can tell.

Experiment	Conclusive	Easy	Cheap	Total	Problem
Disc with weight above	**3**	**3**	**3**	**9**	**Frictional coupling effects?**
GPS satellite Test	**2**	**3**	**3**	**8**	**Might be due to other effects**
Drop Tower Test	**3**	**2**	**2**	**7**	**Might not be detectable**
Redo the Tajmar experiment	2	2	2	6	Hard to duplicate exactly
Redo the Podkletnov experiment	2	2	2	6	Hard to duplicate exactly
Stefan–Boltzmann, low-T	2	2	2	6	Need temperatures of 200 pK!
Galactic resonant rings	2	2	2	6	Might be due to other effects
NEMS & radio waves	1	2	2	5	EM effects to consider
CERN & radio waves	2	1	1	4	Expensive, EM effects

126   *Physics from the Edge: A New Cosmological Model for Inertia*

how conclusive, easy and cheap they would be. For example, the experiment with the rotating disc and the weight above it would be relatively conclusive, and gets a mark of 3, because the experiment can be brought to an equilibrium and then the disc can be spun at different rates and in different directions providing a nice set of independent data that can be compared with the predictions of MiHsC.

On the other hand, the experiment with CERN particles, the disc and NEMS with applied radio waves would not be so conclusive since all sorts of electromagnetic effects would have to be calculated too.

You may very well have a better experimental suggestion, and I would encourage you to try it. Please let me know how it goes.

Chapter 7

MiHsC and Faster Than Light Travel

The conservation of momentum combined with special relativity suggests that an object's inertial mass depends on its velocity v as

$$m_i = \frac{m_0}{\sqrt{1 - \frac{v^2}{c^2}}} \qquad (7.1)$$

where m_0 is the rest mass. By playing around with this equation you can show that when the velocity of an object is small compared to the speed of light ($v \ll c$) the inertial mass is close to the rest mass m_0, but when the speed v approaches the speed of light (c) the inertial mass m_i approaches infinity and therefore further increases of speed are impossible. This effect has been verified in particle accelerators and there are also logical reasons for it (there are restrictions from causality too). This limits the range of interstellar travel within the lifetime of those left behind on Earth.

As suggested in this book, MiHsC also affects inertial mass. Combining special relativity (Eq. (7.1)) and MiHsC (Eq. (4.7)) produces this expression for the inertial mass m_i

$$m_i = m_0 \frac{(1 - \frac{2c^2}{a\Theta})}{(1 - \frac{v^2}{c^2})^{\frac{1}{2}}} \qquad (7.2)$$

where m_0 is the rest mass. Using Newton's second law gives the acceleration

$$a = \frac{F}{m_i} = F \frac{(1 - \frac{v^2}{c^2})^{\frac{1}{2}}}{m_0(1 - \frac{2c^2}{a\Theta})} \qquad (7.3)$$

Multiplying both sides by the denominator and rearranging gives

$$a = \frac{F}{m_0}\left(1 - \frac{v^2}{c^2}\right)^{\frac{1}{2}} + \frac{2c^2}{\Theta} \qquad (7.4)$$

The first term on the right hand side is the usual one from special relativity that states that if $v = c$, then no matter how large a force (F) is applied to an object it will not accelerate. The new result from MiHsC is the second term, which states that even when $v = c$ there will always be a minimum acceleration of: $2c^2/\Theta \sim 6.7 \times 10^{-10}$ m/s^2 even if the force applied (F) is zero. To explain intuitively: as a spacecraft's speed approaches c, special relativity predicts that its inertia increases and the acceleration falls towards zero, but MiHsC says that inertia depends on the existence of Unruh waves and at low accelerations these become too long to fit within the Hubble diameter, so whereas special relativity predicts that the inertia increases to infinity at speed c, MiHsC predicts that as this happens, Unruh-inertia dissipates. The result is that a residual acceleration remains (Eq. (7.4), term 2).

This prediction is supported indirectly by the observations of Perlmutter *et al.* (1999) and Reiss *et al.* (1998) who observed this same value of acceleration for the distant stars which are travelling away at speeds relative to us of close to c. This acceleration has been attributed to arbitrary 'dark energy' but it is predicted by MiHsC.

This suggests a paradox: a star near the observable universe's edge is moving at speed c, a moment later, because of cosmic acceleration, it is moving faster than c. This is contrary to special relativity alone, but not when MiHsC is also considered (Eq. (7.4)).

The minimum acceleration predicted by MiHsC is tiny: from rest, it would take 14 billion years to reach the speed of light (It is interesting that the empirical acceleration coefficient needed by MoND (Milgrom, 1983) is of a similar size to that predicted by MiHsC, and as Milgrom noted: would produce the speed of light in the age of the universe, 13.7 billion years).

Equation 7.4 implies that the way to boost this force-independent MiHsC acceleration and achieve faster than light (FTL) travel in a shorter time is not to increase F in the first term which will run up against the constraints of special relativity, but instead reduce Θ (the Hubble scale) in the second term.

FTL Test 1: Man-Made Event Horizons

Of course, shrinking the observable universe is likely to remain impossible for some time, but it may be possible to create a smaller 'effective' Hubble scale. I suggested in the papers McCulloch (2008a, 2013) that one way to achieve this would be to use the metamaterials recently devised by Pendry *et al.* (2006) and Leonhardt (2006). They have demonstrated that electromagnetic radiation (which forms part of the Unruh radiation) can be bent around an object, which must be smaller than the wavelength, using a metamaterial (a specially designed metal structure), making that object invisible to an observer at that wavelength (cloaking). By instead bending Unruh radiation (or just the EM-component of Unruh radiation) back towards a spacecraft it may be possible to create an event horizon similar to the one assumed by MiHsC to exist at the edge of the observable universe, the size of which is the Θ in Eq. (7.4). The value of Θ is usually 2.7×10^{26} m, but for a spacecraft surrounded by a carefully arranged metamaterial shell, Θ could be the diameter of the shell, making the MiHsC-acceleration predicted by Eq. (7.4) very much larger. The speed of light could then be achieved and passed at a much greater acceleration. For example, if a craft is accelerated at $9.8\,\mathrm{m/s^2}$ the wavelength of Unruh radiation would be

$$\lambda = \frac{4\pi^2 \beta c^2}{a} \sim 7.3 \times 10^{16}\,\mathrm{m} \tag{7.5}$$

If the spacecraft was surrounded by a metamaterial that bent this wavelength around it, then according to MiHsC, the object's inertial mass would reduce. Pendry and Wood (2007) have devised metamaterials that are able to control very low electromagnetic frequencies including wavelengths as large as that above (bending other Unruh fields would be harder).

FTL Test 2: Particle Accelerators

The effects of MiHsC have not been observed in particle accelerators which accelerate particles to close to the speed of light. This is because these particles travel along circular trajectories

130   *Physics from the Edge: A New Cosmological Model for Inertia*

and are therefore highly accelerated, making MiHsC's effects less apparent. This can be shown quantitatively using the combined MiHsC + relativity model for inertia (Eq. (7.2)). Substituting the acceleration of a particle around the CERN particle accelerator: $a = v^2/r$, where r is the radius of the accelerator and v is the velocity, into Eq. (7.2), we get

$$m_i = m_0 \frac{(1 - \frac{2c^2 r}{v^2 \Theta})}{(1 - \frac{v^2}{c^2})^{\frac{1}{2}}} \qquad (7.6)$$

Replacing all the known constants with values, so: $c = 3 \times 10^8$ m/s, $r = 4$ km (for the Large Hadron Collider at CERN), and $\Theta = 2.7 \times 10^{26}$ m, gives

$$m_i = \frac{m_0(1 - 2.6 \times 10^{-6}/v^2)}{\sqrt{1 - 1.1 \times 10^{-17} \times v^2}} \qquad (7.7)$$

Using a binomial series for the denominator: $(1 - \frac{v^2}{c^2})^{\frac{1}{2}} \sim 1 - \frac{v^2}{2c^2} \cdots$ so that approximately

$$m_i \sim \frac{m_0(1 - (2.6 \times 10^{-6}/v^2))}{1 - (5.5 \times 10^{-18} v^2)} \qquad (7.8)$$

The change in the inertial mass from special relativity and MiHsC can now be found. When $v = 0.9c$ the effect of MiHsC is 22 orders of magnitude smaller than the change due to relativity, and when v is higher still, the effect of MiHsC decreases even further. Therefore it would be extremely difficult to detect the effects of MiHsC in a particle accelerator.

However, MiHsC predicts that measureable effects, and FTL, might be achieved for systems with high linear velocity with very low accelerations around them, such as in many of the examples discussed in Chapter 5. Two other possible examples not yet mentioned are cosmic rays entering the atmosphere and galactic axial jets (it is possible that M87 has an axial jet that actually moves faster than light, and not just apparently).

FTL Test 3: Galactic Jets

MiHsC predicts that just as the Earth flyby craft are boosted when they exit along the Earth's polar axis, objects may lose inertial mass and be more easily accelerated along galactic axes. Galactic jets have been known for some years and Biretta *et al.* (1999) observed the M87 axial jet and calculated the apparent speed of recognisable 'knots' of light within the jet, taking account of the estimated distance to M87. They found an apparent speed of $6c$. It was shown by Rees (1966) that this apparent superluminal speed is enhanced by an optical illusion caused by special relativity. From Rees (1966) the apparent speed (v_{app}) of a relativistic object moving at an angle θ to the observer depends on its real speed (v) as

$$v_{app} = \frac{v \sin \theta}{1 - \cos \theta} \tag{7.9}$$

According to Biretta *et al.* (1999) the most likely angle of the M87 jet to our line of sight is 64.5°, and they also said that because of the observed shape of the knots: "placing the jet within 20° of the line of sight presents several challenges".

Table 1 shows the assumed angle (column 1) and the real velocity implied by the observed apparent velocity of $v_{app} = 6c$, using Eq. (7.9). In order to produce real velocities less than the speed of light for the M87 jet, it is necessary to assume unrealistic angles of less than 20°. This of course is a controversial area, and estimates of the jet's angle or the distance to M87 may change, but it is possible that this is evidence for the FTL speeds that MiHsC implies can be achieved.

Table 1.   The assumed angle to the line of sight of the M87 jet and the implied absolute speed.

Assumed angle to line of sight	Absolute velocity
64°	$3.7c$
30°	$1.6c$
20°	$1.06c$

Conclusions

It is obvious to me that a recasting of physics is needed. This can be seen conceptually by saying that quantum mechanics is incompatible with general relativity, so one of them, at least, is wrong. On the observational side, anomalies are piling up, and some of them, like the anomalous dynamics of globular clusters, cannot hope to be explained by dark matter.

Yet, there is still a huge stigma in physics against alternative theories. If you like, the ecosystem has too little diversity to be healthy. It seems that physicists today confronted with a fascinating anomaly will propose invisible matter or invisible dimensions, anything 'hidden' rather than changing Einstein's theories. This is a centuries old fail safe of mankind's: when something is not understood it is attributed to invisible entities, like Ptolemy's epicycles, Descartes' vortices, dark matter or dark energy. These tend to stick around for a long time because they are so flexible, therefore hard to disprove.

In this book I have presented an alternative theory that doesn't rely on over-complex invisible entities, but relies simply on Unruh radiation (which may or may not have been observed, but is at least observable). This theory moves away from the particles and smooth fields used by Newton and Einstein, and is based on a way of thinking that uses horizons and information which modify the Unruh radiation.

I have suggested that when objects accelerate, Rindler horizons form in the opposite direction to their acceleration, suppressing the Unruh radiation in that direction and opposing the acceleration. This correctly models inertia.

I have discussed how for very low accelerations even the far off Hubble horizon starts to cancel the Unruh radiation, so that inertia

134   *Physics from the Edge: A New Cosmological Model for Inertia*

decreases in a new way for these low accelerations, and this difference can explain the observed anomalous galaxy rotation, cosmic acceleration, the low-L CMB anomaly, maybe the flyby anomalies, the Tajmar effect and possibly the Podkletnov results and similar results obtained by Poher and Modanese.

Finally, I have tried to suggest unambiguous tests. Probably the best test would be to produce an acceleration so large (by accelerating spinning discs) that it produces Unruh waves and the inertial mass and dynamics of a nearby mass are detectably affected by it. Another route would be to try to directly interfere with the Unruh waves that make up the inertial mass using electromagnetic waves.

I have shown how this new theory suggests a new way to launch things into space and that it suggests very tentatively that faster than light travel may be possible.

Despite what we are told at university, physics is not set in stone, but is a work in progress. Hopefully these new ideas will encourage some of you to test or develop them further. If they are proven right by experiment, then, not only will we have fascinating new insights into our cosmos, but also new ways to travel through it.

Appendix A

Brief Discussion of Papers on MiHsC

This is a quick chronological summary of all the papers I have written so far on MiHsC (as of 2013).

2007. I assumed that inertia was caused by Unruh radiation, and subject to a Hubble-scale Casimir effect so that some Unruh waves are disallowed because they don't fit within the Hubble scale. This leads to a new loss of inertia for low accelerations. I applied this model (called MiHsC) to the trajectories of the Pioneer spacecraft and showed that the loss of inertia leads to an extra Sunward acceleration equal to the Pioneer anomaly. I remember the delightful comments of the reviewer of this paper who was amused by my use of the word "forecast" instead of prediction (I worked at the UK Meteorological Office at the time) and said something like: "I don't quite believe his solution, but it's more plausible than others that have been published, so..." Subsequent work by Turyshev *et al.* (2012) has proposed that the Pioneer anomaly could be due to an anisotropic radiation of heat, but the model they use is complex with adjustable parameters and there is no decay in the anomaly with time to back a thermal model. *Mon. Not. Roy. Astron. Soc.*, **376**, 338–342, 2007.

2008a. The flyby anomalies are anomalous changes of a few mm/s in the speed of spacecraft flying by the Earth. In this paper I tried to model them by saying that when the craft pass through a zone where the net acceleration is low they lose inertial mass by MiHsC and speed up by momentum conservation. I spent the better part of a year modelling trajectories in my spare time, and it did not work because I did not yet consider mutual accelerations. However, in this paper, I also suggested controlling inertia by bending Unruh waves using metamaterials. *J. Brit. Interplanet. Soc.*, **61**, 373–378, 2008.

2008b. This paper was inspired by observations of Anderson *et al.* (2008) that showed that the flyby anomaly was large when the spacecraft came towards the Earth at the equator and left at the pole. When I downloaded the paper it upset me because I couldn't explain it, but then I realised with joy that I could model it using MiHsC if I considered the 'mutual' accelerations between masses, since the mutual acceleration between a spacecraft and masses in the spinning Earth is lower closer to the spin axis. MiHsC then predicts the craft's inertia is lower near the pole and to conserve momentum the craft speeds up. This models the flyby anomalies fairly well without adjustable parameters, but not perfectly. *Mon. Not. Roy. Astron. Soc. Lett.*, **389**(1), L57–60, 2008.

2010a. In this paper I applied MiHsC to the observations of Tajmar *et al.* (2007) who noticed an unexplained acceleration of accelerometers close to rotating rings. I took the idea of mutual accelerations further and considered the inertial mass of the accelerometer to be dependent (via MiHsC) on not only its acceleration with respect to the spinning ring but to the fixed stars too (with a nod to Ernest Mach). The idea was sound but I messed up the maths. I realised my error the night before I was due to give an important talk on it in Berne! I had to write another paper to correct it (see 2011a). *Europhys. Lett.*, **89**, 19001.

2010b. This was a more detailed look at the prediction by MiHsC that since Unruh waves lengthen as accelerations reduce, and because the Unruh waves cannot in principle be observed if they are greater than the Hubble scale, there must then be a minimum acceleration allowed in nature. I showed that this is close to the observed cosmic acceleration that is usually attributed to arbitrary 'dark energy'. MiHsC also predicts the observed minimum mass for disc galaxies seen by McGaugh *et al.* (2009). In this paper I also suggested modifying the inertial mass of an object by interfering with Unruh radiation using EM radiation. *Europhys. Lett.*, **90**, 29001.

2011a. I corrected my mathematical mistake (in 2010a) and MiHsC worked well but didn't fit one of Tajmar's results. When I emailed Tajmar he told me that particular result was due to a wrong stepper motor, so I was ecstatic. The prediction of MiHsC is that when the ring accelerates the accelerometer gains inertial mass and has to move with the ring to conserve the overall momentum of the system. MiHsC predicts the results very well. This paper and

2010a won "Best of Year" awards from the EPL journal. *Europhys. Lett.*, **95**, 39002.

2011b. This was my attempt to explain the weight loss seen by Podkletnov when he vibrated and span a superconducting disc below various test masses. MiHsC provided a possible explanation, but not a complete one and I couldn't go further because I had no way to know what the accelerations/vibrations of the disc were when it was spun. This paper on a controversial experiment led to me being consigned to gen-ph on the arXiv and led to a couple of critical letters being sent to my university faculty, but then great joy as the head of my school wrote an email supporting my academic freedom. *Physics Procedia*, **20**, 134–139.

2012. I must have submitted nearly six different papers several times each over four years trying to model a disc galaxy with MiHsC with different methods. With each rejection I tried again and my method became simpler till eventually, there was nothing for the reviewers to reject it on. ☺ MiHsC predicts the rotation speeds of dwarf, disc and galaxy clusters within the error bars without any adjustable parameters and most crucially: **without dark matter**. I have yet to model a galaxy in detail though. *Astrophys. Space Sci.*, **342**(2), 575–578.

2013. In all the papers above I used a Hubble-scale Casimir effect to model 'deviations' from standard inertia, and just assumed standard inertia. In this paper I proposed that standard inertia is due to a Rindler-scale Casimir effect. As an object accelerates, say, to the right, a Rindler horizon forms to its left since information further away can never catch up. A Rindler-scale Casimir effect then suppresses Unruh waves on the left, so that the object feels more Unruh radiation pressure from the right. This pressure pushes it back against its acceleration: an elegant model for inertia that needs no adjustable parameters. This model also represents a new way of thinking about motion and energy in terms of horizons and information. *Europhys. Lett.*, **101**, 59001.

Appendix B

Hawking Radiation

To arrive very roughly at Hawking's (1974) result we can apply the uncertainty principle to a photon emitted from the event horizon of a black hole as follows

$$\Delta p \Delta x \sim \hbar$$

The Δp is the uncertainty in momentum of the photon and the Δx is the uncertainty in position. Heisenberg stated that their product was equal to h-bar which is Planck's constant divided by 2π. Since the outwards photon's companion has disappeared into the black hole somewhere, the uncertainty in position is close to the radius of the black hole of mass M which is

$$R = \frac{2GM}{c^2}$$

Putting this into the uncertainty principle for Δx we get

$$\Delta p \sim \frac{\hbar c^2}{2GM}$$

Since the energy $E = pc$ this gives

$$\Delta E = \frac{\hbar c^3}{2GM}$$

When energy is in a random thermal form, it can be written in terms of temperature as $E = kT$. Hence,

$$T = \frac{\hbar c^3}{2GMk}$$

This is the Hawking temperature of a black hole and when this is derived properly an 8π appears in the denominator.

Appendix C

Unruh Radiation

We can now do a similar analysis to Hawking's for the Rindler horizon caused by an object's acceleration (this was done by Unruh in a more rigorous way). Heisenberg's uncertainty principle for momentum (p) and position (x) is

$$\Delta p \Delta x \sim \hbar$$

The uncertainty in space (Δx) of an object with an acceleration a is then determined by the distance to its Rindler horizon, c^2/a:

$$\Delta p \frac{c^2}{a} \sim \hbar$$

So

$$\Delta p \sim \frac{\hbar a}{c^2}$$

Since $E = pc$

$$\Delta E \sim \frac{\hbar a}{c}$$

If the energy is thermalised so $E = kT$ then

$$\Delta T \sim \frac{\hbar a}{ck}$$

Unruh (1976) derived this formula first, after Hawking had done his work on the black hole and the only difference was that he had an extra 2π in the denominator, giving

$$\Delta T \approx \frac{\hbar a}{2\pi ck}$$

What this means is that as an object's acceleration (a) increases the background temperature it measures (T) also increases. This is called Unruh radiation and it originates from the Rindler horizon.

Appendix D

Asymmetric Casimir Effect

This is a derivation of a model for inertia using an asymmetric Casimir effect (for the complete derivation, see McCulloch, 2013). A single particle is considered for simplicity. It is accelerating to the right as in Figure 5 in Chapter 4, reproduced below. The radiation pressure (force) on any small surface area (A) exposed to isotropic (direction independent) radiation is

$$F = \frac{uA}{3}$$

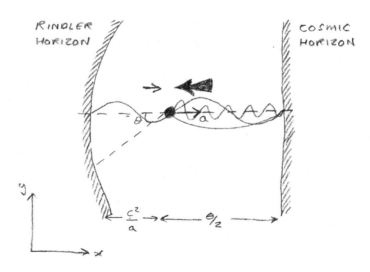

Figure 1.   A schematic showing a particle accelerating to the right (the black circle). The shading shows the Rindler horizon close by to its left (at a distance c^2/a away) and the cosmic horizon far away to its right. The Unruh radiation is damped more on the left side producing an asymmetric Casimir effect (the force indicated by the two unequal opposing arrows) that pushes the particle (O) to the left against its accelerations: a model for inertia.

The three comes from the fact that there are three directions possible (x, y and z) and so the total energy is split between them. The u stands for the Unruh radiation energy density, A is the surface area of the particle intercepting this radiation and this is assumed to be a small part of the surface area of the whole particle. Now we need the net difference between the forces from the right and the left so we draw a line through the particle at an arbitrary angle θ (the dashed line) and calculate the forces along this line (see the arrows in the figure) and take its component along the x-axis by multiplying by $\cos\theta$.

$$dF_x = \frac{u_{\text{left}} A \cos\theta}{3} - \frac{u_{\text{right}} A \cos\theta}{3}$$

The energy content of the Unruh radiation coming from the right (the second term on the right hand side) is subject to the usual Hubble-scale Casimir effect, so $u_{\text{right}} = u(1 - \lambda/4\Theta)$, where u is the unmodified energy and λ is the peak wavelength of the Unruh radiation spectrum. In contrast, the energy of the Unruh radiation coming from the left (the first term on the right hand side) is subject to a Rindler-scale Casimir effect and this horizon is much closer, at a distance of $c^2/a\cos\theta$ (where $a\cos\theta$ is the component of the acceleration in the direction of θ) so that $u_{\text{left}} = u(1 - \lambda/4(c^2/a\cos\theta))$. The difference, the net force in the x-direction, is

$$dF_x = \frac{uA\cos\theta}{3}\left(1 - \frac{\lambda a \cos\theta}{4c^2} - 1 + \frac{\lambda}{4\Theta}\right)$$

We can cancel the $1 - 1$ in the brackets, so this simplifies to

$$dF_x = \frac{u\lambda A \cos\theta}{3}\left(\frac{1}{4\Theta} - \frac{a\cos\theta}{4c^2}\right)$$

We can now integrate the contribution from all the possible angles. To do this we integrate within the plane of the page, from the angle $\theta = 0$ to $\pi/2(90°)$, then double this to get the result for the whole x–y plane and then integrate this circularly 180° or π around the x-axis (around the angle ϕ) to calculate the total force:

$$dF_x = 2 \times \frac{u\lambda A}{3} \int_0^\pi \int_0^{\pi/2} \left(\frac{\cos\theta}{4\Theta} - \frac{a\cos^2\theta}{4c^2} \right) d\theta d\phi$$

$$dF_x = \frac{2u\lambda A}{3} \int_0^\pi \left[\frac{\sin\theta}{4\Theta} - \frac{a\theta}{8c^2} - \frac{a\sin 2\theta}{16c^2} \right]_0^{\pi/2} d\phi$$

which is

$$dF_x = \pi \times \frac{2u\lambda A}{3} \left[\frac{1}{4\Theta} - \frac{\pi a}{16c^2} \right].$$

The first term in the brackets is the MiHsC correction to what we normally expect from inertia. It is tiny compared to the second term unless the acceleration, a, is very small and the second term becomes small. If we assume a normal (terrestrial) acceleration then we can neglect this first term, so

$$dF_x = -\frac{\pi^2 u\lambda Aa}{24c^2}$$

Since u is the energy within a volume V, so $u = E/V = hc/\lambda V$, then

$$dF_x = -\frac{\pi^2 hAa}{24cV}$$

This is the predicted force on one spherical particle, due to the asymmetric Casimir effect. It means that the Rindler horizon that appears in the accelerated particle's reference frame damps Unruh radiation in the direction opposite to its acceleration, so the radiation there is unable to balance the momentum imparted by the radiation hitting it from the other direction. The result is a net force counter to the acceleration (the minus sign) and this looks just like inertia.

The ratio A/V = area/volume could be simplified by choosing a distance scale (x) smaller than the particle so that $A/V = x^2/x^3 = 1/x$. Using the Planck distance ($l_P = 1.616 \times 10^{-35}$ m) as x for

146   *Physics from the Edge: A New Cosmological Model for Inertia*

example, then the equation above becomes

$$dF_x = -\frac{\pi^2 ha}{24cl_P}$$

And the inertial mass for the Planck scale particle is 5.5×10^{-8} kg which is about twice the Planck mass. It was later kindly pointed out to me by J. Giné that I had made a factor of 2 error and the 24 in this result should be 48. This makes the prediction better: only 26% greater than the Planck mass. A paper to correct this (Giné and McCulloch, 2014) is being reviewed at the time of writing this book.

References

Adams, D., 1979. *The Hitch-Hiker's Guide to the Galaxy*. Pan Books Ltd.

Adler, R., 2006. Gravity. In *The New Physics for the Twenty-First Century*. G. Fraser (Ed.). Cambridge University Press.

Anderson, J.D., P.A. Laing, E.L. Lau, A.S. Liu, M.M. Nieto and S.G. Turyshev, 1998. *Phys. Rev. Lett.*, **81**, 2858.

Anderson, J.D., P.A. Laing, E.L. Lau, A.S. Liu, M.M. Nieto and S.G. Turyshev, 2002. *Phys. Rev. D.*, **65**, 082004.

Anderson, J.D., J.K. Campbell, J.E. Ekelund, J. Ellis and J.F. Jordan, 2008. *Phys. Rev. Lett.*, **100**, 091102.

Antreasian, P.G. and J.R. Guinn, 1998. *AIAA/AAS Astrodynamics Specialist Conference and Exhibition*, No. 98-4287, doi:10.2514/6.1998-4287.

Akhmedov, E.T. and D. Singleton, 2007. *JETP Lett.*, **86**, 615.

Aristotle, 330BC. *Physics*, translated by R.P. Hardie and R.K. Gaye.

Beversluis, M.R., A. Bouhelier and L. Novotny, 2003. Continuum generation from single gold nanostructures through near-field mediated intraband transitions. *Phys. Rev. B*, **68**, 115433.

Boersma, S.L., 1996. A maritime analogy of the Casimir effect. *American Journal of Physics*, **64**, 539–540.

Bradley, J., 1971. *Mach's Philosophy of Science*. The Athlone Press, University of London.

Brooks, M., 2009. *13 Things That Do Not Make Sense*. Profile Books Ltd.

Casimir, H., 1948. On the attraction between two perfectly conducting plates. *Proc. Con. Nederland. Akad. Wetensch.*, **B51**, 793.

Cook, N., 2001. *The Hunt for Zero Point*. Arrow Publishers.

Crum, L., 1994. Sonoluminescence. *Physics Today*, **47**(9), 22.

Davies, P.C.W., 1975. Scalar production in Schwarzschild and Rindler metrics. *J. Phys. A.*, **8**(4), 609.

Dicke, R.H. and P.J.E. Peebles, 1979. The big bang cosmology — enigmas and nostrums. In *General Relativity: An Einstein Centenary Survey*. S.W. Hawking and W. Israel (Eds.). Cambridge University Press.

Ehrenberg, I.M., S.E. Sarma and B.-I. Wi, 2012. A three-dimensional self-supporting low loss microwave lens with a negative refractive index. *J. Appl. Phys.*, **112**, 073114.

Einstein, A., 1905. On the electrodynamics of moving bodies. *Annalen der Physik*, **17**, 891.

148   *Physics from the Edge: A New Cosmological Model for Inertia*

Farington, B., 1961. *Greek Science*. Pelican Books.

Feynman, R.P., R.B. Leighton and M. Sands, 1977. *Lectures on Physics: Volume 1.*

Frank, P., 1969. Einstein, Mach and logical positivism. In *Albert Einstein, Philosopher-Scientist: The Library of Living Philosophers Volume VII.* Paul Arthur Schilpp (Ed.), p. 272.

Fraser, G., (Ed.), 2006. *The New Physics for the Twenty-First Century.* Cambridge University Press.

Freedman, W.L., 2001. Final results of the Hubble space telescope key project to measure the Hubble constant. *Astrophys. J.*, **553**, 47–72.

Freedman, W.L. and E.W. Kolb, 2006. Cosmology. In *The New Physics for the Twenty-First Century.* G. Fraser (Ed.). Cambridge University Press.

Fulling, S.A., 1973. Nonuniqueness of canonical field quantization in Riemannian spacetime. *Phys. Rev. D.*, **7**(10), 2850.

Giné, J. and M.E. McCulloch, 2014. On the origin of inertial mass. Submitted for publication.

Gleick, J., 2003. *Isaac Newton*. Harper Collins.

Graham, R.D., R.B. Hurst, R.J. Thirkettle, C.H. Rowe and P.H. Butler, 2008. *Physica C*, **468**, 383.

Guth, A.H., 1981. The inflationary universe: a possible solution to the horizon and flatness problems. *Phys. Rev. D*, **23**, 347.

Hafele, J.C. and R.E. Keating, 1972. Around-the-World Atomic Clocks: Observed Relativistic Time Gains. *Science*, **177**(4044), 168–170.

Hawking, S., 1974. Black hole explosions. *Nature*, **248**, 30.

Hoffmann, B., 1906. *Relativity and Its Roots*. W.H. Freeman and Co.

Hoyle, F., 1948. A new model for the expanding universe. *Mon. Not. Roy. Astron. Soc.*, **108**, 372–383.

Hulse, R.A. and J.H. Taylor, 1975. Discovery of a pulsar in a binary system. *Astrophys. J.*, **195**, L51–L53.

Knuuttila, T.A., 2000. DSc thesis. Helsinki University of Technology. http:lib.tkk.fi/Diss/2000/isbn9512252147/.

Mannheim, P.D., 1990. *Gen. Relativ. Gravit.*, **22**, 289.

McCulloch, M.E., 2007. Modelling the Pioneer anomaly as modified inertia. *Mon. Not. Roy. Astron. Soc.*, **376**, 338–342.

McCulloch, M.E., 2008a. Can the flyby anomalies be explained using a modification of inertia? *J. Brit. Interplanet. Soc.*, **61**, 373.

McCulloch, M.E., 2008b. Modelling the flyby anomalies using modified inertia. *Mon. Not. Roy. Astron. Soc. Lett.*, **389**, L57.

McCulloch, M.E., 2010a. Can the Tajmar effect be explained using a modification of inertia? *Europhys. Lett.*, **89**, 19001.

McCulloch, M.E., 2010b. Minimum accelerations from quantised inertia. *Europhys. Lett.*, **90**, 29001.

McCulloch, M.E., 2011a. The Tajmar effect from quantised inertia. *Europhys. Lett.*, **95**, 39002.

McCulloch, M.E., 2011b. Can the Podkletnov effect be explained by quantised inertia? *Physics Procedia*, **20**, 134–139.

McCulloch, M.E., 2012. Testing quantised inertia on galactic scales. *Astrophys. Space Sci.*, **342**(2), 575–578.

McCulloch, M.E., 2013. Inertia from an asymmetric Casimir effect. *Europhys. Lett.*, **101**, 59001.

McCulloch, M.E., 2014. A toy cosmology from a Hubble-scale Casimir effect. *Galaxies*, **2**, 81–88.

McGaugh, S.S., J.M. Schombert, W.J.G. de Blok and M.J. Zagursky, 2009. *Astrophys. J.*, **708**, L14.

Meyer, J.C. *et al.*, 2005. *Science*, **309**(5740), 1539–1541.

Michelson, A.A. and E.W. Morley, 1887. *Am. J. Sci.*, **34**, 333.

Milgrom, M., 1983. A modification of the Newtonian dynamics as a possible alternative to the hidden mass hypothesis. *Astrophys. J.*, **270**, 365–370.

Milgrom, M., 1998. The modified dynamics as a vacuum effect. Arxiv: 9805346v2.

Milgrom, M., 2005. In *Mass Profiles and Shapes of Cosmological Structure*. G. Mamon, F. Combes, C. Deffayet and B. Fort (Eds.). EAS Publications Series Vol. 20, EDP Sciences, Les Ulix Cedex A, p. 217.

Minovitch, M., 1961. A method for determining interplanetary free-fall reconnaissance trajectories. *JPL Technical Memo.*, TM-312-130, pp. 38–44. 23rd August, 1961.

NASA, 2005. NASA Facts: Gravity Probe B Official Fact Sheet. http://einstein.stanford.edu/content/fact sheet/GPB FactSheet-0405.pdf.

NASA, 2012. On the expansion of the universe. NASA Glenn research centre. Accessed, 12.11.2012.

Ohanian, H.C., 1985. *Physics*. W.W. Norton & Co. Ltd.

Pais, A., 1982. *Subtle is the Lord: The Science and the Life of Albert Einstein*. Oxford University Press.

Penrose, R., 2004. *The Road to Reality*. Jonathon Cape.

Podkletnov, E.E. and R. Nieminen, 1992. A possibility of gravitational shielding by bulk $YBa_2Cu_3O_{7-x}$ superconductor. *Physica C*, **203**, 441–444.

Podkletnov, E.E., 1997. Weak gravitational shielding properties of composite bulk $YBa_2Cu_3O_{7-x}$ superconductor below 70 K under e.m. field. Arxiv: cond-mat/9701074v3.

Rindler, W., 2001. *Relativity, Special, General and Cosmological*. Oxford University Press.

Sarton, G., 1993. *Ancient Science Through the Golden Age of Greece*. Couier Dover, p. 248.

Scarpa, R., G. Marconi and R. Gilmozzi, 2006. Globular clusters as a test for gravity in the weak acceleration regime. Arxiv: 0601581v1.

Schwinger, J., 1992. *Proc. Natl. Acad. Sci. USA*, **89**, 4091–4093; 11118–11120.

Smolin, L., 2006. *The Trouble with Physics: The Rise of String Theory, the Fall of Science and What Comes Next*. Penguin books.

Smolyaninov, I.I., 2008. Photoluminescence from a gold nanotip in an accelerated reference frame. Arxiv: cond-mat/0510743.

Stanford University, accessed in 2013. Testing Einstein's Universe. http://einstein.stanford.edu/MISSION/mission1.html.

Tajmar, M., F. Plesescu, B. Seifert and K. Marhold, 2007. *AIP Conf. Proc.*, **880**, 1071.

150   *Physics from the Edge: A New Cosmological Model for Inertia*

Tajmar, M., F. Plesescu, B. Seifert, R. Schnitzer and I. Vasiljevich, 2008. *AIP Conf. Proc.*, **69**, 1080.

Tajmar, M., F. Plesescu and B. Seifert, 2009. *J. Phys. Conf. Ser.*, **150**, 032101.

Taylor, J.H. and J.M. Weisberg, 1982. A new test of general relativity: gravitational radiation and binary pulsar PSR 1913 + 16. *Astrophys. J.*, **253**, 908–920.

Tully, R.B. and J.R. Fisher, 1977. A new method of determining distances to galaxies. *Astron. & Astrophys.*, **54**(3), 661–673.

Turyshev, S.G., V.T. Toth, G. Kinsella, S.-C. Lee, S.M. Lok and J. Ellis, 2012. Support for the thermal origin of the Pioneer anomaly. *Phys. Rev. Lett.*, **108**, 241101.

Unzicker, A., 2008. Why do we still believe in Newton's laws? Facts, myths and methods in gravitational physics. Arxiv: 0702009v2.

White, M., 2007. *Galileo: Antichrist*. Pheonix books.

Will, C.M., 2001. The confrontation between general relativity and experiment. *Living Reviews in Relativity*, **4**, 4.

Wolfe, J., 2012. Flute Acoustics. University of New South Wales, Australia. http://www.phys.unsw.edu.au/jw/fluteacousics.html#airjet.

Ziebart, M., A. Sibthorpe, P. Cross, Y. Bar-Sever and B. Haines, 2007. Cracking the GPS-SLR orbit anomaly. *Proceedings of ION-GNSS-2007*. Fort Worth, Session F-4, pp. 2033–2038.

Zyga, L., 2012. Dark matter effect might be explained by modified way to calculate inertial mass. *Phys. Org*, Sept 18th, 2012.

Index

152   *Physics from the Edge: A New Cosmological Model for Inertia*

Printed in the United States
By Bookmasters